AIN AMKO

Topics in Advanced Emission Control and Diagnostic Sensors

SP-1180

All SAE papers, standards, and selected
books are abstracted and indexed in the
Global Mobility Database.

Published by:
Society of Automotive Engineers, Inc.
400 Commonwealth Drive
Warrendale, PA 15096-0001
USA
Phone: (412) 776-4841
Fax: (412) 776-5760
May 1996

Permission to photocopy for internal or personal use, or the internal or personal use of specific clients, is granted by SAE for libraries and other users registered with the Copyright Clearance Center (CCC), provided that the base fee of $7.00 per article is paid directly to CCC, 222 Rosewood Drive, Danvers, MA 01923. Special requests should be addressed to the SAE Publications Group. 1-56091-823-3/96$7.00.

Any part of this publication authored solely by one or more U.S. Government employees in the course of their employment is considered to be in the public domain, and is not subject to this copyright.

No part of this publication may be reproduced in any form, in an electronic retrieval system or otherwise, without the prior written permission of the publisher.

ISBN 1-56091-823-3
SAE/SP-96/1180
Library of Congress Catalog Card Number: 96-68115
Copyright 1996 Society of Automotive Engineers, Inc.

Positions and opinions advanced in this paper are those of the author(s) and not necessarily those of SAE. The author is solely responsible for the content of the paper. A process is available by which discussions will be printed with the paper if it is published in SAE Transactions. For permission to publish this paper in full or in part, contact the SAE Publications Group.

Persons wishing to submit papers to be considered for presentation or publication through SAE should send the manuscript or a 300 word abstract of a proposed manuscript to: Secretary, Engineering Meetings Board, SAE.

Printed in USA

PREFACE

Automotive engineers are being challenged on several fronts. On one hand, the California TLEV, LEV, and ULEV standards and the European Stage III standards are requiring lower and lower emissions from cars and trucks. At the same time, the OBD II legislation requires that vehicles be able to self-diagnose several components on board which can have an effect on the emissions from the vehicle. Finally, manufacturers are being encouraged to produce cars and trucks which consume less fuel. Engineers and scientists throughout the world are working on various phases of these demanding requirements. The result of this work should be a fleet of fuel-efficient low-emitting vehicles which satisfies the needs for dependable transportation and environmental conservation.

This SAE special publication, <u>Topics in Advanced Emission Control and Diagnostic Sensors</u> (SP-1180), provides an excellent resource for engineers and scientists new to the area of automotive emission research as well as for individuals who have been involved in the area for a considerable amount of time. I would like to thank all of the authors of the papers as well as the reviewers for their efforts in making this book and this session a success.

Joseph R. Theis
Delphi Energy and Engine
Management Systems

Session Organizer

TABLE OF CONTENTS

961129 **Advanced Catalyst Studies of Diesel NOx Reduction for Heavy-Duty Diesel Trucks** 1
    M. Kawanami, A. Okumura, and M. Horiuchi
        Nippon Shokubai Co., Ltd.
    A. Schäfer-Sindlinger
        Degussa Japan Co., Ltd.
    K. Zerafa
        ICT Inc.

961130 **New Type of NOx Sensors for Automobiles** 11
    Yukio Nakanouchi, Hideyuki Kurosawa, Masaharu Hasei, Yongtie Yan, and Akira Kunimoto
        Riken Corp.

961131 **Rare Earth Catalysts for Purification of Auto Exhaust** 19
    Zhang Zhiqiang and Fang Dachun
        Dongfeng Motor Corporation Auto Industry Institute
    Zhi Rentao and Shang Chengjia
        Beijing Science and Technology Univ.
    Zhang Zhongtai and Huang Yong
        Tsinghua Univ.

961133 **Changes in Pollutant Emissions from Passenger Cars Under Cold Start Conditions** 23
    Robert Joumard and Robert Vidon
        Institut National de Recherche sur les Transports et leur Sécurité
    Laurent Paturel
        Savoy Univ.
    Gérard de Soete
        Institut Français du Pétrole

961134 **Applications and Benefits of Catalytic Converter Thermal Management** 35
    Steven D. Burch, Matthew A. Keyser, Chris P. Colucci, Thomas F. Potter, and David K. Benson
        National Renewable Energy Lab.
    John P. Biel
        Benteler Industries, Inc.

961137 **Electrically Heated Catalyst - Design and Operation Requirements** 41
    F. Terres, H. Weltens, and D. Froese
        Heinrich Gillet GmbH & Co. KG

961129

Advanced Catalyst Studies of Diesel NOx Reduction for Heavy-Duty Diesel Trucks

M. Kawanami, A. Okumura, and M. Horiuchi
Nippon Shokubai Co., Ltd.

A. Schäfer-Sindlinger
Degussa Japan Co., Ltd.

K. Zerafa
ICT Inc.

Copyright 1996 Society of Automotive Engineers, Inc.

1. ABSTRACT

New catalysts with HC (hydrocarbon) storage ability to improve NOx conversion and to minimize fuel penalty over the US Heavy Duty Transient cycle were developed. Without secondary fuel addition, simultaneous reduction of 13% NOx and about 30% particulate was achieved by storing HC from the engine during low temperature portions of the transient cycle and releasing and using the stored HC for NOx conversion at higher temperatures. With only 1% secondary fuel addition, NOx reduction can be increased to 25%, and the particulate conversion remained relatively constant at about 20%. More than 30% NOx reduction can be obtained with 3% fuel penalty. All the pollutants (NOx, PM, HC and CO) were reduced with 0-1% secondary fuel addition.

2. INTRODUCTION

It is widely recognized that NOx and particulate matter are harmful pollutants that can cause serious environmental damage. Therefore, their reduction in many countries is demanded (1). One of the major sources of NOx and particulate matter is diesel engines. Therefore, tighter standards for particulate and NOx emissions from diesel vehicles are proposed in many regions of the world. Oxidation catalysts with sulfate suppression ability are widely known to be effective in reducing the particulate matter in diesel exhaust (2-5). Many engine manufactures have started to use oxidation catalysts that can reduce HC, CO and the SOF (soluble organic fraction) portion of the particulate matter, which consists of unburned diesel fuel and lubricating oil, to meet the '94 US Heavy Duty regulations. However, the '98 US Heavy Duty regulations will require an additional 20% NOx reduction. In addition, the 2004 statement of principals is targeting another 50% reduction in NOx.

Recently, many research activities have focused on catalysts that can reduce NOx in diesel exhaust, including selective NOx reduction systems that use HC as the reductant or NOx decomposition systems that use Cu/ZSM-5 catalysts (6-9). NOx reduction or NOx absorption-decomposition systems with precious metal catalysts are also suggested (10-12). However, the information available on the applicability of these catalysts to diesel engine exhaust where the particulate reduction function is also a requirement, is rather limited. Only a few papers report on simultaneous reduction of NOx and particulate (13-15). In our previous paper (15), we have demonstrated simultaneous reduction of 12% NOx and 25% particulate matter over the US Heavy Duty Transient cycle with a 2-3% fuel penalty. However, more NOx conversion and less fuel penalty are desired for practical applications.

In this paper, we report our progress in developing new catalysts. First, we screened washcoat materials for HC storage ability. Then, NOx performance and sulfate suppression ability were improved. Finally, the NOx reduction temperature window was shifted to 300-450°C to match the temperatures at which most of the NOx is emitted during the US Heavy Duty Transient cycle. The NOx conversion and HC storage analysis results from the US Heavy Duty Transient tests using this newly developed catalyst are also reported.

3. EXPERIMENTAL

CATALYSTS - The catalysts used for the experiments were all prepared at our laboratories, and were of the washcoated cordierite monolith type, with 400 cells per square inch. A list of the catalysts tested is given in Table 1.

Table 1. Experimental catalysts tested.

Catalyst A	Cu/ZSM-5
Catalyst B	Precious Metal Type
Catalyst C	Base Metal Type
Catalyst D	Base Metal Type with HC storage ability
Catalyst E	Base Metal Type with HC storage ability
Catalyst F	Base Metal Type with HC storage ability
Catalyst G	Base Metal Type with HC storage ability

MODEL GAS TESTS - The model gas was prepared by gas-mixing equipment and bottled gases, and introduced to a temperature controlled reactor containing catalyst bricks of 2.4 cm diameter and 6.6 cm length. The model gas feed was controlled in order to produce a space velocity of $50,000h^{-1}$. The model gas composition is given in Table 2.

Table 2. Model Gas Composition.

HC (C_3H_6)	2400 ppmC_1
NO	270 vppm
CO	350 vppm
SO_2	50 vppm
O_2	6.0 vol%
CO_2	10.7 vol%
H_2O	10 vol%
N_2	balance

HC ADSORPTION TESTS - Five kinds of materials were washcoated. The substrates were cordierite monoliths with 400 cells per square inch. Diesel fuel was vaporized, and introduced to a temperature controlled reactor containing catalyst bricks of 5.3 cm diameter and 12.7cm length. The gas feed was controlled to result in a space velocity of $20,000h^{-1}$. The diesel fuel concentration in the air was monitored by a FID HC analyzer. The HC concentration was adjusted to 100ppmC_1. The amount of HC adsorbed was calculated by integrating the HC adsorption peak. The HC adsorption speed was calculated from the amount of HC adsorbed during the first 2 minutes of the tests.

DIESEL ENGINE TESTS - A Caterpillar 3116, 6.6 liter intercooled turbocharged diesel engine was used for the diesel engine experiments. The catalyst inlet temperature was controlled by operating the engine at 1400rpm while varying the engine torque during the steady-state tests. The US Heavy Duty Transient test cycle was simulated by using an ONO SOKKI engine control system HU-2010. The transient tests were performed twice for each catalyst and only the second test results were utilized.

Two pieces of 5.66"x6.0" catalyst bricks were used in a parallel configuration during steady-state tests resulting in a space velocity of $55,000h^{-1}$. Two pieces of 7.5"x7.0" catalysts were used in series during the transient cycles.

Secondary fuel was delivered to the exhaust by using a dual plunger-type pump, and was sprayed into the exhaust gas stream with air. Secondary fuel was added according to 2 different patterns: one is constant fuel addition during the transient cycle (Constant), and the other is fuel addition during the highest NOx generation portions of the transient cycle (Pattern 2). The injection profiles are presented in Figure 6. The sulfur content of the diesel fuel used for the engine and secondary HC injection was 0.03 wt%.

HC, CO and NOx concentrations were measured by

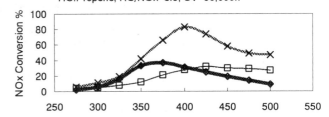

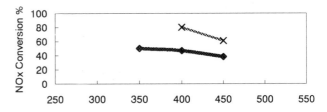

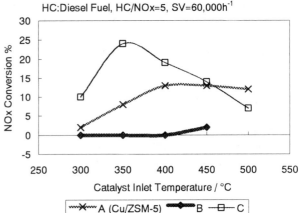

Fig.1. NOx Conversions of Catalyst A through C under Three Different Gas Conditions.

a HORIBA MEXA9100D system. Particulate matter was measured by diluting all the exhaust gas with air, using a secondary full dilution tunnel during the transient tests. The comprehensive test procedures for the particulate measurements are described in the previous paper (15).

4. RESULTS AND DISCUSSION
4.1 CATALYST DEVELOPMENT

Our target is to develop catalysts which have high NOx activity under diesel exhaust conditions. Diesel fuel, which is very different from propene used in the model gas tests, is the only practical HC source under diesel exhaust conditions. So we tested the effects of gas source on the NOx reduction performance, prior to the development of new catalysts. For this test, Cu-ZSM-5 (Catalyst A), catalysts B and C were tested using three different gas sources: model gas

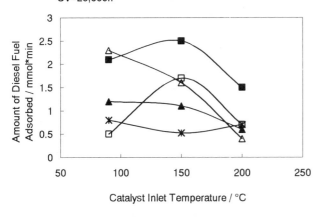

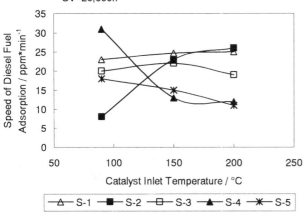

Fig.2. The Amount and Speed of Diesel Fuel Adsorption when Washcoat Material S-1 through S-5 are used.

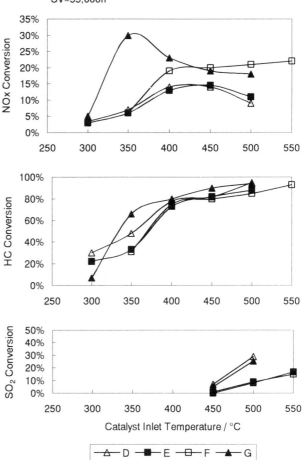

Fig.3. NOx, HC and SO_2 Conversions of Catalyst D through Catalyst G.

(propene as the HC), exhaust gas from a lean-burn gasoline engine (propene and ethene as the major HC), and exhaust gas from a diesel engine (diesel fuel as the HC). The NOx reduction performance under these three different conditions is shown in Fig.1. The results of model gas tests and lean-burn engine tests have good correlation. This is probably due to the HC species being similar in both tests. (Most of HC species in lean-burn gasoline engines are unsaturated hydrocarbons such as propene and ethene.) However, the results obtained under diesel engine conditions are different from those obtained under model gas and lean-burn gasoline conditions. The NOx reduction performance of a catalyst is greatly influenced by the gas source. This is due to the different activities of catalysts with different HC species in the gas. Therefore, diesel NOx catalysts were evaluated under diesel engine conditions in our laboratory.

Since we cannot expect much HC from diesel engines, it is necessary to utilize all the HC from the engine for NOx reduction to improve NOx conversion without HC addition. It is difficult to reduce NOx at temperatures below 200°C because even precious metal catalysts are not active below 200°C. So, HC emitted from the engine during low temperature portions of the transient cycle (150-200°C) should be stored in the catalyst and utilized for NOx reduction at higher temperatures. Therefore, we developed new catalysts in the following way. First, we screened washcoat materials for HC storage ability. Then, NOx performance and sulfate suppression ability were improved. Finally, the NOx reduction temperature window was shifted to 300-450°C to match the temperatures at which most of the NOx is emitted during the US Heavy Duty Transient cycle.

First, washcoat materials with high HC storage ability were screened for this purpose. Diesel fuel was used as the HC, since the major part of the HC from diesel engines is unburned diesel fuel. The ability of washcoat materials to adsorb HC (diesel fuel) is shown in Fig.2. Among the five materials tested, washcoat material S-2 had the highest diesel fuel adsorption ability (amount and speed) in the temperature range from 150 to 200°C. We used this washcoat material for the preparation of catalyst D, to add HC storage ability.

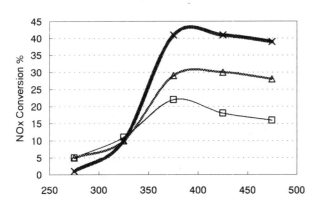

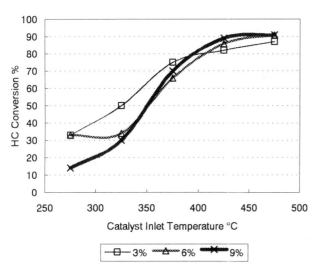

Fig.4 NOx and HC Conversions of Catalyst G with 3 to 9% Secondary Fuel.

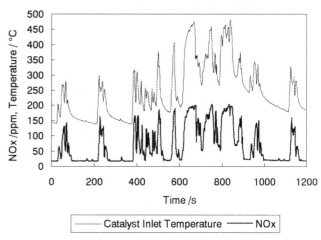

Fig.5. Catalyst Inlet Temperature and NOx Concentration during the Transient Cycle.

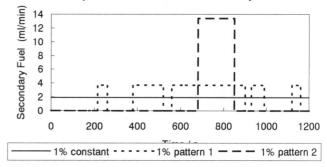

Fig.6. The Injection Patterns of Secondary Fuel during the Transient Cycle while Keeping Total Amount of Secondary Fuel Added at 1% of Engine Fuel Consumption.

The next step was to improve NOx performance and sulfate suppression ability of this catalyst. The results are shown in Fig.3. We tried to suppress SO_2 activity of catalyst D. The resulting catalyst E shows less than 10% SO_2 conversion at 500°C while keeping the NOx activity the same. Catalyst F was developed to increase NOx activity without increasing SO_2 activity. Finally, catalyst G was developed to shift the NOx reduction temperature window of catalyst F from 400-550 to 300-450°C. Catalyst G shows the highest NOx conversion among the four HC storage type catalysts. The relationship between NOx conversion, secondary fuel injection amount and catalyst inlet temperature of catalyst G is shown in Fig.4. The amount of secondary fuel was increased with the increase of the catalyst inlet temperature (engine load) to keep fuel penalty constant. With 3% secondary fuel, 22% NOx reduction is observed. More than 40% NOx conversion was observed with 9% secondary fuel.

In this way, we have developed a new catalyst which has high NOx conversion, sulfate suppression ability and HC storage ability. Also, the new catalyst has NOx reduction temperature window of 300-450°C. This temperature range matches the temperatures at which most of the NOx is emitted during the US Heavy Duty Transient cycle. The superior performance of catalyst G prompted us to test this catalyst under the US Heavy Duty Transient cycle.

4.2 CATALYST PERFORMANCE USING THE US HEAVY DUTY TRANSIENT CYCLE

Catalyst inlet gas temperatures and engine-out NOx concentrations during the US Heavy Duty Transient cycle are shown in Fig.5. One potential method to achieve higher NOx conversion is a combination of a precious metal and a base metal catalyst with secondary fuel injection Pattern 1, and the other potential method is to use only a base metal catalyst with secondary fuel injection Pattern 2. (Refer to Fig.6)

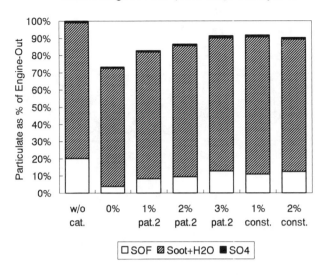

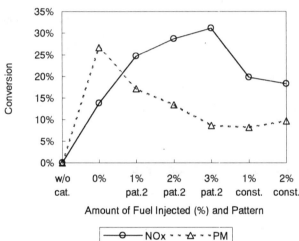

Fig.7. Conversions and Particulate Matter during the Transient Cycle with Secondary Fuel Injection. Catalyst : G

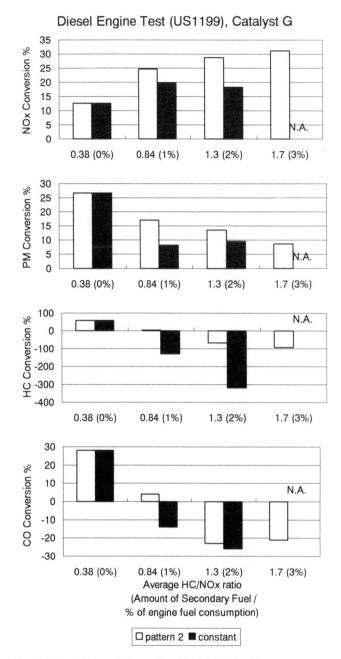

Fig.8. The Effects of Secondary Fuel Addition Patterns on Conversions of NOx, PM, HC and CO.

According to Pattern 2, large amounts of secondary fuel is injected only at temperatures from 300 to 450°C, when most of the NOx is emitted and where a base metal catalyst is active. For detailed discussions on these secondary fuel addition patterns, please refer to our previous paper (15).

For newly developed catalyst G, the latter method with secondary fuel injection Pattern 2 should be applied because catalyst G is a base metal catalyst. (Refer to Table 1.) Different amounts of secondary fuel were injected using either constant injection or Pattern 2 over the transient cycle, and the more effective injection pattern for simultaneous NOx and particulate matter reduction was evaluated.

The results are shown in Fig.7. It can be seen that 25% NOx and 17% particulate matter reduction was achieved simultaneously with only a 1% secondary fuel addition by Pattern 2. Even without secondary fuel addition, simultaneous reduction of 13% NOx and 27% particulate matter was achieved. By increasing the secondary fuel addition to 3%, NOx conversion increased to 32% and the particulate matter conversion decreased to 8%. Catalyst G sufficiently oxidizes SOF, and sulfate formation is suppressed.

For catalyst G, the results of different secondary fuel addition patterns are compared in Fig.8. It is obvious that the conversions of all the pollutants (NOx, PM, HC and CO) are improved when secondary fuel is added according to Pattern 2 rather than by constant rate secondary fuel addition, while the total amount of secondary fuel added is kept constant. Note that in addition to NOx and particulate, HC and CO are also reduced. All the pollutants are reduced from raw emission

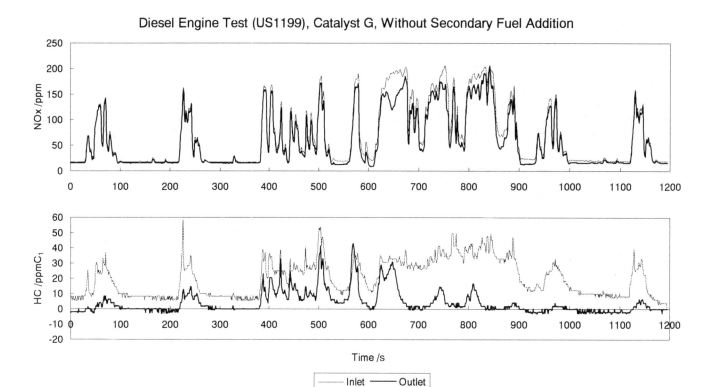

Fig.9. Second-by-second Data of NOx and HC Concentrations during the Transient Cycle without Secondary Fuel Addition.

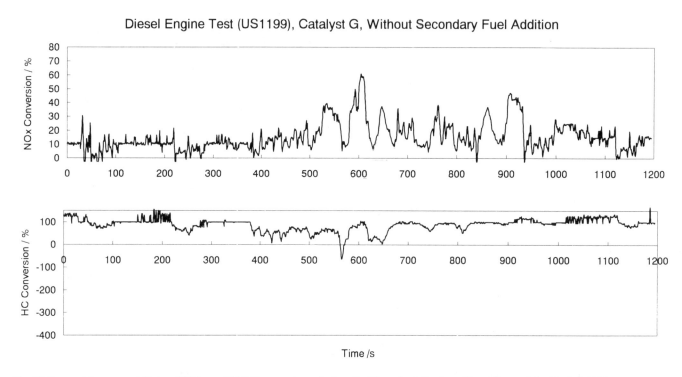

Fig.10. Second-by-second Data of NOx and HC Conversions during the Transient Cycle without Secondary Fuel Addition.

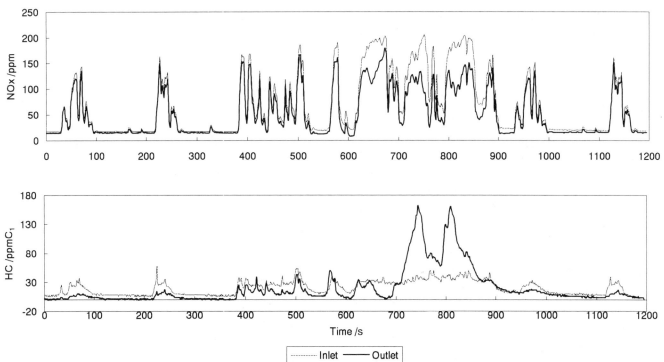

Fig.11. Second-by-second Data of NOx and HC Concentrations during the Transient Cycle with 1% Secondary Fuel Addition According to Pattern 2.

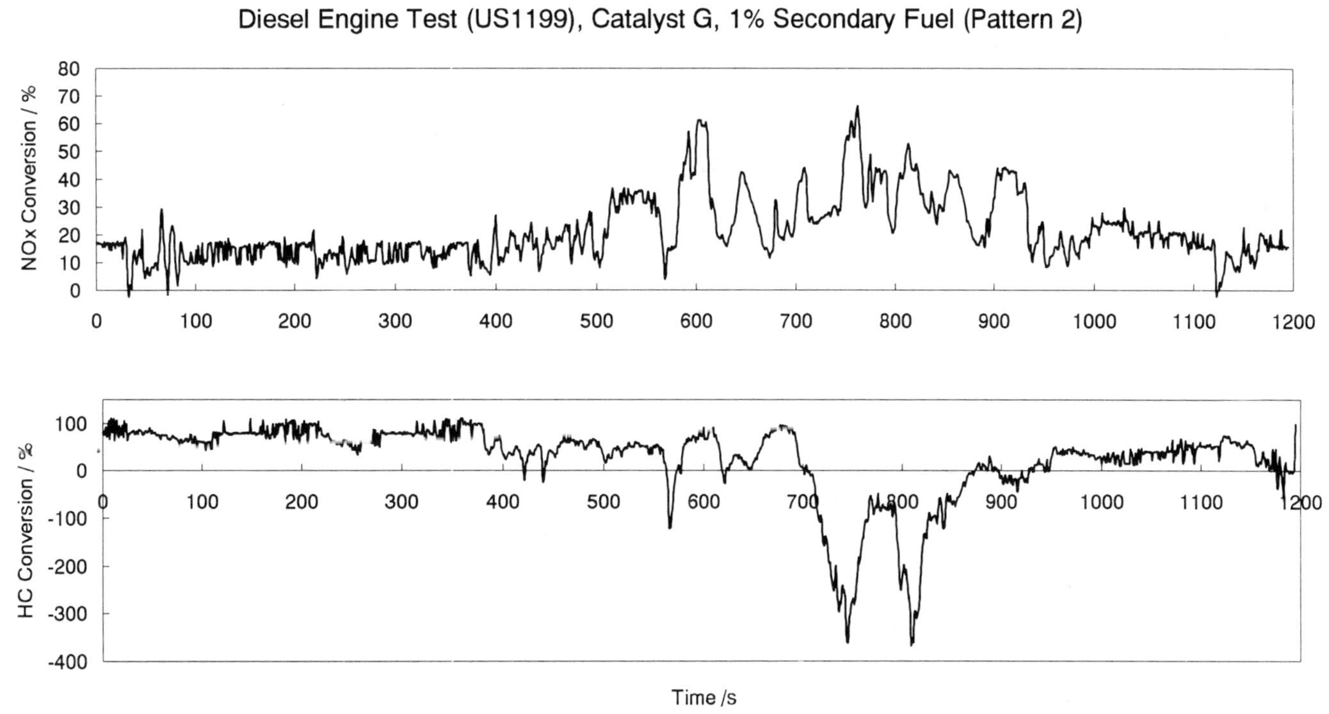

Fig.12. Second-by-second Data of NOx and HC Conversions during the Transient Cycle with 1% Secondary Fuel Addition According to Pattern 2.

levels when 0-1% secondary fuel is added according to Pattern 2. That is, catalyst G functioned as a 4-way catalyst (NOx, PM, HC and CO) under these conditions.

Catalyst G shows very good results for simultaneous NOx and HC reduction during the US Heavy Duty Transient tests. The NOx reduction temperature range of catalyst G fits best with the catalyst inlet temperature during the transient tests. The NOx activity of catalyst G is the highest in the temperature range of 300-450°C.

4.3 NOx CONVERSION AND HC OXIDATION ANALYSIS DURING THE TRANSIENT CYCLE

To analyze the periods during the transient cycle when NOx conversion occurs, second-by-second data for HC and NOx concentration before and after catalyst G were obtained when no secondary fuel was added (Fig.9). The average HC/NOx ratio was 0.38. NOx and HC conversions calculated from the data in Fig.9 are shown in Fig.10. Second-by-second data when 1% secondary fuel is added according to Pattern 2 are shown in Figs.11 and 12. Note the inlet HC concentration in Fig.11 shows the HC concentration before secondary fuel addition.

As seen from Fig.9, no NOx reduction is observed during 0-350 seconds of the transient cycle, but most of the HC is reduced. The catalyst inlet temperatures are 150-250°C during this period (Fig.5), and oxidation of HC does not occur at these low temperatures with catalyst G. Therefore, HC is being stored in the catalyst. The catalyst inlet temperatures increase to 200-300°C at 400-600 seconds and the stored HC is gradually released, and some NOx reduction is observed. The release is particularly evident at 570 second where HC conversion goes negative. At 600-900 seconds, catalyst inlet temperatures increase to 250-450°C (Fig.5) and the NOx catalyst is fully activated, where 10-40% NOx reduction is observed (Fig.10). Especially at 650 seconds, much of the HC is released from the catalyst and NOx conversion increases to 40%. At 650-900 seconds, stored HC is completely consumed and about 10% NOx conversion is observed. At 900-1200 seconds, NOx conversion gradually decreases with a decrease in the catalyst inlet temperature, and the catalyst starts to store HC again. In the case of steady-state engine tests, all the catalysts showed maximum NOx conversion activity at temperatures where 50 to 70% HC conversion is observed. However, the relationship between HC conversion and NOx conversion during the transient cycle tests is more difficult to determine, due to the HC "storage" effect (Fig.10).

The results when 1% secondary fuel is added according to Pattern 2 are similar to the results without secondary fuel addition, except at 680-850 seconds. The average HC/NOx ratio was 0.84. As seen from Figs.11 and 12, 20-60% NOx reduction is observed at 680-850 seconds, during the time that the secondary fuel is added. The secondary fuel addition starts just after the complete consumption of stored HC in the catalyst, and contributes to NOx reduction in the transient cycle effectively.

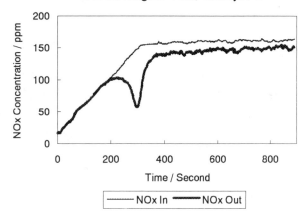
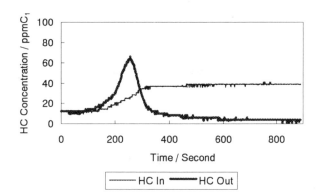
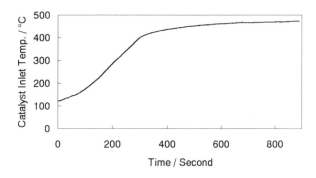

Fig.13. NOx and HC Concentrations and Catalyst Inlet Temperatures when Increasing Engine Speed and Torque at Constant Rates after the Transient Tests.

4.4 HC STORAGE ANALYSIS

Engine speed and torque was increased from idle to 1800rpm 70kg·m at constant rates and NOx and HC levels before and after catalyst G were monitored to check the HC "storage" assumption. This test was performed after the transient tests, to allow the catalyst to store HC. Increasing engine speed and load increases the catalyst temperature and space velocity, so the stored HC should be released by this test if the HC "storage" assumption is correct. The results are shown in Fig.13. It is seen from Fig.13 that catalyst inlet temperature increases with the increase of speed and torque of the engine. At catalyst inlet temperatures from 300 to 400° C, tailpipe HC levels exceed engine-out HC levels, and more than 60% NOx conversion is observed at that time. It is obvious that the catalyst stored HC at low temperatures and released it in the temperature range of 300-400° C. The temperature range for releasing the HC fits catalyst G's NOx reduction temperature window. This HC storage and release characteristic of the catalyst contributed to NOx reduction over the transient cycle, especially when no secondary fuel is added. We added secondary fuel to increase HC/NOx ratio during the US heavy duty transient cycle to improve the NOx conversion. However, with proper engine modifications (in order to increase engine out HC/NOx ratios), it may be feasible to reduce more than 30% of the NOx without adding secondary fuel.

5. CONCLUSIONS

We have shown that catalyst development and secondary fuel addition strategy is expected to play a large role in the NOx reduction for heavy duty diesel engines for on-highway vehicles. It is very important to develop catalysts with NOx reduction temperature windows which match exhaust gas temperatures to maximize NOx performance.

The results of our experiments can be summarized as follows:

1. A hydrocarbon storage function in the catalyst has shown to improve NOx conversion over the US Heavy Duty Transient cycle with minimum fuel penalty.
2. Without secondary fuel addition, simultaneous reduction of 13% NOx and about 30% particulate was achieved.
3. With only 1% secondary fuel addition, NOx reduction increased to 25%, and particulate conversion remained relatively constant at about 20%.
4. More than 30% NOx reduction can be obtained with 3% fuel penalty.
5. All the pollutants (NOx, PM, HC and CO) are reduced with 0-1% secondary fuel addition.

With proper engine modifications (in order to increase engine out HC/NOx ratios), more than 30% NOx conversion may be feasible without secondary fuel addition.

6. REFERENCES

1. M.P.Walsh, "Global Trends in Diesel Particulate Control, 1995 Update", SAE 950149.
2. I.Fukano, K.Sugawara, K.Sasaki, T.Honjou, and S.Hatano, "A Diesel Oxidation Catalyst for Exhaust Emissions Reduction", SAE 932958.
3. M.Horiuchi, K.Saito, and S. Ichihara, "The Effects of Flow-through Type Oxidation Catalysts on the Particulate Reduction of 1990's Diesel Engines", SAE 900600.
4. J.A.Dystrup, W.H.Lane, J.P.Timmons, and A.L.Smith, "The Development of a Production Qualified Catalytic Converter" SAE 930133.
5. D.Standt and A.Koenig, "Performance of Zeolite-Based Diesel Catalysts", SAE950749.
6. M.Konno, T.Chikahisa, T.Murayama, and M.Iwamoto, "Catalytic Reduction of NOx in Actual Diesel Engine Exhaust", SAE 920091.
7. D.R.Monroe, C.L.DiMaggio, D.D.Beck, and F.A.Matekunas, "Evaluation of a Cu/Zeolite to Remove NOx from Lean Exhaust", SAE 930737.
8. G.Muramatsu, A.Abe, M.Furuyama, and K,Yoshida, "Catalytic Reduction of NOx in Diesel Exhaust", SAE 930135.
9. M.J.Heimrich and Marvin L.Deviney, "Lean NOx Catalyst Evaluation and Characterization", SAE 930736.
10. R.Burch, P.J.Millington, and A.P.Walker, "Mechanism of the Selective Reduction of Nitrogen Monoxide on Platinum-based Catalysts in the Presence of Excess Oxygen", Applied Catalysts B: Environmental 4 (1994) p.65-94.
11. B.H.Engler, J.Leyrer, E.S.Lox, and K.Ostgathe, "Catalytic Reduction of Nitrogen Oxides in Diesel Exhaust Gas", paper presented at the International Congress on Catalysis and Automotive Pollution Control, CAPOC III, Brussels, April 1994.
12. N.Miyoshi, K.Ishibashi, T.Tanizawa, R.Muramoto, S.Takeshima, T.Tanaka, S.Matsumoto, M.Saiki, H.Doi, and S.Tateishi, "NOx Storage Reduction Catalyst for Lean-burn Engines", Proceedings of 74th Annual Meeting A of Catalysis Society of Japan, Kagoshima, (1994) p.98-101.
13. K.C.C.Kharas and J.R.Theis, "Performance Demonstration of a Precious Metal Lean NOx Catalysts in Native Diesel Exhaust", SAE950751.
14. J.Leyrer, E.S.Lox and W.Strehlau, "Design Aspects of Lean NOx Catalysts for Gasoline and Diesel Engine Applications", SAE952495.
15. M.Kawanami, M.Horiuchi, J.Leyrer, E.Lox, and D.Psaras, "Advanced Catalyst Studies of Diesel NOx Reduction for On-Highway Trucks", SAE950154.

961130

New Type of NOx Sensors for Automobiles

Yukio Nakanouchi, Hideyuki Kurosawa, Masaharu Hasei, Yongtie Yan, and Akira Kunimoto
Riken Corp.

Copyright 1996 Society of Automotive Engineers, Inc.

ABSTRACT

New types of potentiometric NOx sensors suitable for use on automobiles were developed by using stabilized zirconia as a base solid electrolyte. It was found that the sensor with sensing electrodes of metal oxides ($CdMn_2O_4$ or $NiCr_2O_4$) showed excellent response to NOx in the concentrations between 20 and 4000 ppm at temperatures higher than 600 °C. The electromotive force of the sensors was almost linear to the logarithm of the NOx concentrations with positive slope for NO_2 and negative slope for NO. Especially, the sensor fitted with $CdMn_2O_4$ gave excellent responses to NO at 600 °C, while the sensor fitted with $NiCr_2O_4$ showed high sensitive to NO_2 at 650 °C. The sensors were insensitive to CO, CO_2, CH_4, C_3H_6, O_2 and water vapor. The sensors were fabricated as planar types with the reference electrode exposed to the sample gas, so that the sensors were simple in structure and easy to manufacture.

INTRODUCTION

Air pollution from the exhaust of automobiles has been causing a serious problem in our modern society, and the regulations on the emission standards of pollutants from automobile are becoming more severe year by year. The studies about controllability of combustion for gasoline or diesel engines and the reduction of emissions by catalytic converters have been carried out to meet the emission standards [1-5]. The OBD-II regulation from the U.S. Environmental Protection Agency (E. P. A.) now requires whether the catalyst used to clean the exhaust is in order. The monitoring system has been introduced by using the dual oxygen sensors [6][7], although this is not a direct method. If the combustion control and the catalyst monitoring are carried out with a direct detection of NOx (NO and NO_2) concentration, they will become more accurate and effective. For example, if the fuel ignition, EGR rate and so on are controlled with feedback information on NOx concentration in the exhaust, the engine-out emissions may be reduced. In addition, the deterioration of the catalyst can be easily judged by a NOx sensor behind the converter. From these reasons, a NOx sensor suitable for automobiles is urgently needed. The sensor needs to be compact, inexpensive, superior in thermal and chemical stability, and be able to in-situ and continuously detect NOx concentration at elevated temperatures.

While the engine-out NOx from a stoichiometric engine is mostly NO, the exhaust from diesel engines can contain significant amounts of NO_2 (up to 30 % of the NOx [8]). Also, catalytic converters can oxidize some of the NO to NO_2, particularly under lean conditions. Therefore, the ideal sensor would respond equally to NO and NO_2. Alternatively, a sensor sensitive to only NO could be used with a sensor sensitive to only NO_2 in order to determine the total NOx.

So far there have been many reports on NOx sensors based on semiconductive oxides such as In_2O_3-SnO_2 [9], Cr_2O_3-Nb_2O_5 [10], WO_3 [11]. Almost all of these sensors were limited to temperatures below 500 ℃, although the sensing material themselves were superior in thermal stability. On the other hand, the solid electrolyte NOx sensors using Na^+ conductor such as in NASICON ($Na_3Zr_2Si_2PO_{12}$) or β/β''-alumina and a nitrate or nitrite auxiliary phase have been investigated [12][13]. The electromotive force of these sensors were linear to the logarithm of NO or NO_2 concentrations. We have also tried to develop the tubular sensor using stabilized zirconia and a binary nitrate sensing auxiliary phase [14], because stabilized zirconia is quite tough chemically and mechanically and has been extensively utilized for the A/F sensors equipped in automobiles. The tubular sensor was found to show good response to NOx. Furthermore, we have introduced the planar type sensor with the reference electrode exposed to the sample gas [15]. The planar type sensor gave the same NOx sensing behavior as the tubular one, and the dependency of electromotive force on oxygen concentration was mitigated considerably. Though the sensors based on solid electrolytes exhibited the Nernst's response to NOx, these sensors could only be operated below 500 ℃, because of the melting point limit of the nitrate or nitrite. Because of this, the NOx sensors reported up to now can not be used for combustion control for which the working temperature of the sensor is often required to be higher than 500 ℃.

Because of this situation, we attempted to explore a new type of NOx sensor operative at elevated temperature by combining a stabilized zirconia and a metal oxide. As a result, the sensors using Y_2O_3-stabilized zirconia and $CdMn_2O_4$ or $NiCr_2O_4$ gave excellent sensing properties to NO at 600 ℃ or NO_2 at 650 ℃, respectively. This paper deals with the NOx sensing characteristics and mechanisms of the novel sensors.

EXPERMENTAL

A schematic view of the fabricated NOx sensing device is shown in figure 1. An 8 mol% Y_2O_3-stabilized zirconia (YSZ) plate (4×4×0.3 mm) was used as a based solid electrolyte. For a sensing electrode, a metal oxide film (1×2.2 mm) was deposited on the YSZ plate by R.F. magnetron sputtering (JOEL, JEC-SP360R), and a Pt layer was prepared on top of the oxide film by a screen printing method. A Pt reference electrode was prepared on the same plate by the screen printing method, leaving a 1.5 mm-space between the two electrodes. Then the device was annealed in air for 1 hour at 1050 ℃. Pt wires were welded to the reference electrode and the Pt layer on top of the oxide film by electric welding (Unitek, UNIBOND-Ⅱ).

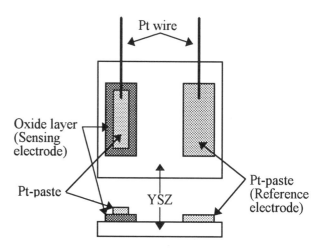

Figure 1. Schematic diagram of zirconia-based NOx sensing device using oxide electrode.

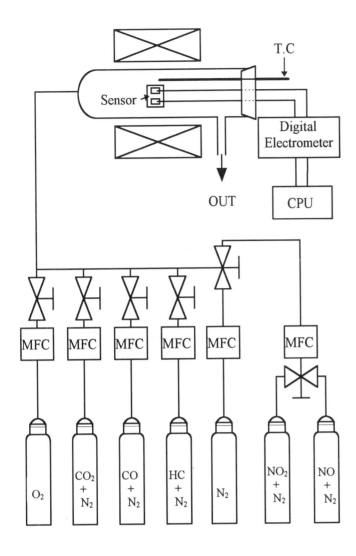

Figure 2. Measuring system for the sensing characteristics of NOx sensor.

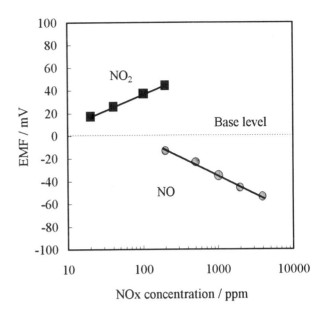

Figure 3. Dependence of EMF on NOx concentration for the sensor fitted with $CdMn_2O_4$ at 600 °C.

The NOx sensing performance was examined in a gas-flow system equipped with a furnace, as shown in figure 2. The sample gases containing various concentrations of NO or NO_2 under a constant oxygen concentration of 4% were prepared by diluting parent gases with nitrogen and oxygen gases, with the total flow rate of 500 cm³/min. The concentrations of CO, CO_2, CH_4 or C_3H_6 were varied by the same method. The electromotive force (EMF) of the device was monitored with a digital electrometer (Advantest, TR8652) while changing the concentrations of various gases at temperatures between 500 °C and 700 °C.

RESULTS AND DISCUSSION

NO SENSOR -   Figure 3 shows the EMF response of a device utilizing a $CdMn_2O_4$ layer upon exposure to several concentrations of NO_2 and NO at 600 °C. The EMF was almost linear to the logarithm of NO or NO_2 concentrations. The EMF decreased with increasing NO concentration in the range 200~4000 ppm, with the slope of -31.8 mV/decade. The EMF increased with increasing NO_2 concentration in the range 20~400 ppm, with the slope of 27.3 mV/decade. The sensor is highly sensitive to both NO and NO_2 at 600 °C. The slope was slightly larger with NO than with NO_2. However, when compared at the same concentration, the sensitivity to NO was smaller than that to NO_2, where the sensitivity is defined as the difference between the EMF with the sample gas and the base level EMF, which usually approaches 0 V. In spite of this, the sensor will be expected to detect the NO concentration in the exhaust before the catalyst for a stoichiometric engine, where the NOx is almost entirely NO. Besides NOx, various

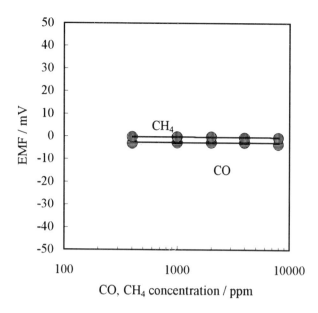

Figure 4. Dependence of EMF on concentration of CO or CH$_4$ for the sensor fitted with CdMn$_2$O$_4$ at 600 °C.

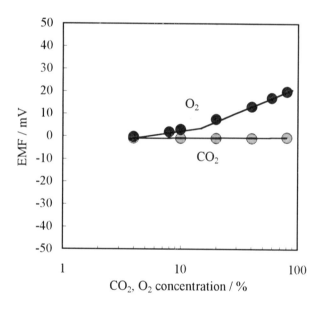

Figure 5. Dependence of EMF on concentration of CO$_2$ or O$_2$ for the sensor fitted with CdMn$_2$O$_4$ at 600 °C.

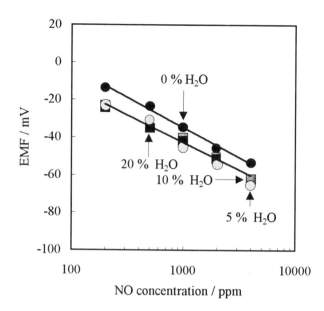

Figure 6. Influence of coexistent water vapor on NO detection for the sensor fitted with CdMn$_2$O$_4$ at 600 °C.

Clearly, the EMF was not noticeably affected by introducing CO or CH$_4$ in the concentration range 400～8000 ppm. Figure 5 gives the shift of EMF with increasing the concentration of CO$_2$ or O$_2$. The sensor was insensitive to CO$_2$. In the case of oxygen, the EMF of the sensor was almost constant in the concentration range between 4 and 10 %, and increased with increasing oxygen concentration in the range 20～80%. Though the oxygen concentration varies with the condition of combustion, it is usually smaller than 10 % with stoichiometric or lean burn engines. Thus, for practical use, the influence of oxygen on the EMF of the sensor is not a serious matter.

The influence of water vapor on NO response was examined because water vapor is also present in the exhaust. Figure 6 shows the influence of coexistent water vapor on the EMF to NO for the sensor. It is found that the EMF of the sensor decreased slightly by introducing water vapor. However, the slopes for the dependence of the EMF on NO concentration remained

gases such as CO, CO$_2$ and hydrocarbons exist in the exhaust from automobiles, and oxygen concentration changes with the condition of combustion, so for practical use the sensor should be resistant to interference from these gases. Figure 4 shows the EMF dependence of the sensor on the concentrations of CO or CH$_4$ at 600 °C.

unchanged in the water-vapor concentration of 0~20 %. So the sensor showed high resistance to water vapor, and can be used to detect NO in the exhaust because the concentration of water vapor is always lower than 20% there.

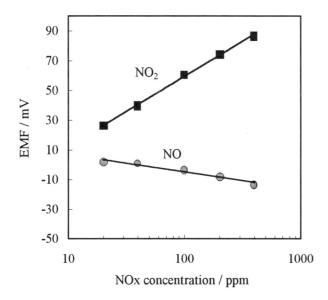

Figure 7. Dependence of EMF on NOx concentration for the sensor fitted with $NiCr_2O_4$ at 650 °C.

NO_2 SENSOR - Figure 7 shows the dependence of the EMF for the device attached with $NiCr_2O_4$ on the NO_2 and NO concentrations at 650 °C. The EMF values were linear to the logarithm of NO_2 and NO concentrations in the range 20~400 ppm, with slopes of 48.5 mV/decade for NO_2 and -13.9 mV/decade for NO, respectively. The slope and sensitivity to NO_2 were much larger than those for NO, though the EMF responses of the sensor to NO_2 and NO were in the opposite direction. These properties of the sensor were examined in the temperature range between 500 °C and 700 °C. The largest slope occurred at 650 °C. Figure 8 shows response transients to NO_2 for the sensor with $NiCr_2O_4$ at 650 °C. The EMF was stable at a fixed concentration and responded quickly to a change in concentration. The times for the 90 % response and recovery was shorter than 15 s. This time included the gas-exchange time in the measuring system, which is estimated to be 8 s. The sensor with high sensitivity to NO_2 can be used to judge the deterioration of the de-NOx catalyst mounted on exhaust manifold. Because the NO_2

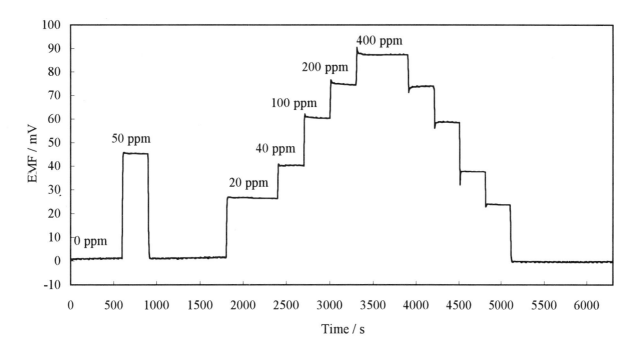

Figure 8. Response transients to NO_2 for the sensor fitted with $NiCr_2O_4$ at 650 °C.

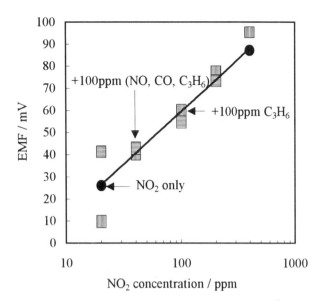

Figure 9. Influence of coexistent gases on NO_2 detection for the sensor with $NiCr_2O_4$ at 650 °C.

will go up for a deteriorated de-NOx catalyst, where the lean environment would promote the oxidation of NO to NO_2. In addition, for a de-NOx system with hydrocarbon injection, the amount of hydrocarbon injection can be optimized by the feedback control with the sensor.

The influences of CO, CO_2, CH_4, C_3H_6 and O_2 on the EMF of the sensor were evaluated at 650 °C, and the results revealed that the EMF was undisturbed by the presence of CO, CO_2, CH_4 up to 1000 ppm or O_2 up to 40%. The EMF of the sensor was changed by introducing C_3H_6, though the slope for the dependence of the EMF on C_3H_6 concentration was smaller than that in the case of NO. Furthermore, the influences of coexistent NO and C_3H_6 on NO_2-response were examined. Figure 9 gives the results obtained at 650 °C. The EMF of the sensor to NO_2 was hardly affected by the coexistent 100 ppm CO, 100 ppm NO and 100 ppm C_3H_6, where the NO_2 concentration was higher than 40 ppm. Like the sensor fitted with $CdMn_2O_4$, the sensor also exhibited high resistance to water vapor.

MECHANISM OF NOx SENSING - The present sensor is constructed in the following cell structure.

Sample gas, Pt | Stabilized zirconia | Metal oxide, Pt, Sample gas
(Reference electrode)|(O^{2-} conductor)|(Sensing electrode)

Here the Pt layer on the metal oxide simply acts as a reservoir of electrons kept in equilibrium with the metal oxide. The potential of the Pt reference electrode, despite the fact that the electrode was exposed to the sample gas, is essentially determined by the O_2 concentration in the sample gas, because the Pt electrode is insensitive to other gases at elevated temperatures. On the other hand, the mechanism of potential generation on the oxide sensing electrode attached on a stabilized zirconia is somewhat complicated. The mechanism based on the mixed potential is widely accepted now, though further investigation is needed. The mixed potential has been discussed on the polarization curve of the sensing electrode in the cases of H_2 sensor and H_2S sensor [16][17]. Here a similar discussion is simply carried out without polarization curve measurements. At the sensing electrode, NOx and O_2 can undergo electrode reactions at the three-phase contact. In the gas containing NO, one can assume cathodic and anodic reactions involving oxide ions as follows.

$$O_2 + 4e^- \rightarrow 2O^{2-} \quad (1)$$
$$2NO + 2O^{2-} \rightarrow 2NO_2 + 4e^- \quad (2)$$

On the other hand, in the gas containing NO_2, one would tend to dissociate down to its equilibrium concentration just opposite to the above.

$$2O^{2-} \rightarrow O_2 + 4e^- \quad (3)$$
$$2NO_2 + 4e^- \rightarrow 2NO + 2O^{2-} \quad (4)$$

The reaction of (1) + (2) or (3) + (4), form a local cell and determines the mixed potential of sensing electrode with the cathodic and anodic reactions. The EMF of the sensor is the deference between the mixed potential of the sensing electrode and the potential of the reference electrode. Clearly, at a fixed concentration of O_2 the EMF increases with increasing NO_2 concentration, and decreases with increasing NO concentration.

SUMMARY AND CONCLUSIONS

New types of potentiometric NOx sensors were fabricated by using stabilized zirconia and metal oxide. The sensors exhibited good sensing performances to NOx at above 600 ℃. The electromotive force of the sensor was almost linear to the logarithm of NOx concentration with positive slope for NO_2 and negative slope for NO. Especially, the sensor fitted with $CdMn_2O_4$ gave excellent responses to NO at 600 ℃, while the sensor fitted with $NiCr_2O_4$ showed high sensitive to NO_2 at 650 ℃. Furthermore, the sensors are insensitive to CO, CO_2, CH_4, C_3H_6, O_2 and water vapor, and exhibited quick responses to NOx, with the time for 90 % response being shorter than 15 s.

The planar sensor with a reference electrode exposed to the sample gas is compact, and all of the constituent materials are superior in thermal stability. These features enhance the possibility of mounting the sensor directly on the exhaust manifold. Now the new sensor capable of self heating is being fabricated and tested. In the near future the sensor will be mounted on exhaust manifold, and its characteristics and durability will be examined for practical use.

REFERENCES

1. S. Shundoh, N. Komori and K. Tsujimura, "NOx Reduction from Diesel Combustion Using Pilot Injection with Pressure Fuel Injection", SAE paper 920461.
2. T. Inoue, S. Matsushita, K. Nakanishi and H. Okano, "Toyota lean combustion system the third generation system", SAE paper 930873.
3. N. Uchida, Y. Daisho, T. Saito and H. Sugano, "Combined effects of EGR and Supercharging on Diesel Combustion and Emissions", ASE paper 930601.
4. M. Konno, T. Chikahisa, T. Murayama and M. Iwamoto, "Catalytic Reduction of NOx in Actual Diesel Engine Exhaust", SAE paper 920091.
5. G. Muramatsu, A. Abe, M. Furuyama and K. Yoshida "Catalytic reduction of NOx in Diesel Exhaust", SAE paper 930135.
6. J. W. Koupal, M. A. Sabourin and W. B. Clemmens, "Detection of Catalyst Failure On-Vehicle Using the Dual Oxygen Sensor Method", SAE paper 910561.
7. J. S. Hepburn, D. A. Dobson, C. P. Hubbard, S. O. Guldberg, E. Thanasiu, W. L. Watkins, B. D. Burns and H. S. Gandi, "A Review of the Dual EGO Sensor Method for OBD-II Catalyst Efficiency Monitoring", SAE paper 942057.
8. J. C. Hilliard and R. W. Wheeler, "Nitrogen Dioxide in Engine Exhaust", SAE paper 790691.
9. G. Sberveglieri and S. Groppelli, "Radio Frequency Magnetron Sputtering Growth and Characterization of Indium-Tin Oxide(ITO) Then films for NO_2 Gas Sensors", Sensors and Actuators, 15 p235-242, (1988).
10. T. Ishihara, K. Shiokawa, K. Eguchi and H. Arai, "Selective Detection of Nitrogen Monoxide by the Mixed Oxide of Cr_2O_3-Nb_2O_5", Chemistry Letters, p997-1000, (1988).
11. M. Akiyama, Z. Zhang, J. Tamaki, N. Miura and N. Yamazoe, "Tungsten Oxide-based Semiconductor Sensor for Detection of Nitrogen Oxides in Combustion Exhaust", Sensors and Actuators B, 13-14 p-619-920, (1993).
12. Y. Shimizu, Y. Okamoto, S. Yao, N. Miura and N.

Yamazoe "Solid Electrolyte NO$_2$ Sensors Fitted with Sodium Nitrate and/or Barium Nitrate Electrodes", Denki Kagaku, 59, p-465-472, (1991).

13. S. Yao, Y. Shimizu, N. Miura and N. Yamozoe, "Using of Nitrate Auxiliary Electrode for Solid Electrolyte Sensor to Detect Nitrogen Oxides", Chemistry Letters, p587-590, (1992).

14. H. Kurosawa, Y. Yan, N. Miura and N. Yamazoe, "Stabilized Zircinia-Based Potentiometric Sensor for Nitrogen Oxides", Chemistry Letters, p1733-1736, (1994).

15. H. Kurosawa, Y. Yan, N. Miura and N.Yamazoe, "Stabilized Zirconia-Based Nox Sensor Operative at High Temperature", Solid Ionics, 79 p338-343, (1995).

16. N.Miura and N. Yamazoe, Edited by T. Seiyama, Chemical Sensor Technology, 1 p123-139, Kodansha / Elsevier, Japan, (1988).

17. Y. Yan, N. Miura and N. Yamazoe, "Potentiometric Sensor Using Stabilized Zirconia and Tungsten Oxide for Hydrogen Sulfide", Chemistry Letters, p1753-1756, (1994).

961131

Rare Earth Catalysts for Purification of Auto Exhaust

Zhang Zhiqiang and Fang Dachun
Dongfeng Motor Corporation Auto Industry Institute

Zhi Rentao and Shang Chengjia
Beijing Science and Technology Univ.

Zhang Zhongtai and Huang Yong
Tsinghua Univ.

Copyright 1996 Society of Automotive Engineers, Inc.

ABSTRACT

The influence of the catalyst loading, the active components, and the size of the pellets on the conversion were discussed. Over a CeCuM'M" catalyst at $25000h^{-1}$ the CO could be oxidized about 53.4% at 150°C. The radial distribution of the rare earth Ce, and the transitional metals M' and M" were homogeneous, but Cu increased gradually from the centre to outside of the pellets.

INTRODUCTION

Automobile exhaust pollution has been a worldwide environmental hazard for many years. Hence regulations were established in various parts of the world to restrict the auto pollutants emission. The United States issued its first federal law "Clean Air Act" for auto emission in 1970s. More stringent limits were gradually adopted in the following years in Japan and in many other countries. This led to the use of catalytic converters. In the early 1980s, the Three-way catalytic converter came into wide use. It allowed the three main auto pollutants to be efficiently reduced simultaneously by means of electronically-controlled fuel injection, oxygen sensors and precise control of the A/F ratio at 14.7.

With the economy booming, China has become one of the fastest growing auto markets over the past several years. As a result, auto pollutants contribute heavily to air pollution, especially in large cities with heavy traffic. On Nov. 8,1993 China promulgated the revised State Standard named "Emission Standard for Pollutants at Idle Speed from the Road Vehicle with a petrol engine." How to eliminate environmental pollution by automotive exhaust has become a problem to be solved promptly. However, the use of precious metal catalysts in three-way catalytic converters is expensive, so this method is difficult to use extensively in China. Therefore, we hope to find the rare earth elements which have been reported to be very active for restricting auto pollutant emission as a substitute for precious metals. Beside the advantages of low cost, thermal stability, high activity of oxidation, reduction and durability, the RE catalysts have the special property of resistance to lead and sulphur poisoning.[5-8]

EXPERIMENTAL

PREPARATION OF SUPPORT-γ-Al_2O_3 power--grind--sifting--mixing with stabilizer and pore making agent--seed making--pelletized(in pellets machine)--dried-- calcined--γ-Al_2O_3 pellets[9]
Main Composition: γ-Al_2O_3(XRD method) Crushing Strength: 8Kg/grain
Specific Surface Area: $120m^2/g$ (BET method) Bulk Specific gravity: 0.7-0.8
Particle Size: 3-4mm, 7-8mm.

Preparation of Catalyst.

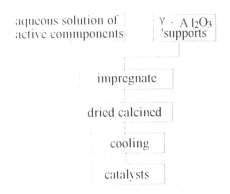

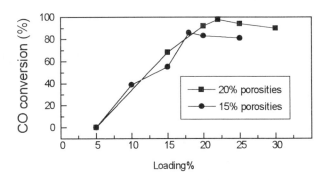

Fig.1 Effect of different loading amount of catalysts

MEASUREMENT OF HC AND CO CONVERSION-The efficiency of the catalyst was determined by an infra-red analyzer of model QGS-04. The catalyst was sealed in the middle of the heated zone with steel wool in a vertical tubular furnace. The reactant gas containing 6600ppm CO and 4500ppm C_3H_8 was passed through the catalysts at space velocities of $25000h^{-1}$ and $5000h^{-1}$. Then, after elevation of the temperature, we recorded and calculated the conversion.

MICROSTRUCTURE OBSERVATION OF CATALYSTS-The surface and distributions of Al, Ce, Cu, Mn, Zr were observed by a scanning electric microscope(SEM:S-450,Hitachi Ltd.,Tokyo).

RESULTS AND DISCUSSION

THE EFFECT OF CATALYST LOADING-Fig.1 shows the effect of CeCuM'M" varying the catalyst loading on the conversion at $25000h^{-1}$[1,2]. The conversion increased with the loading of the catalyst when the support contained 20% porosity. This phenomena can also be viewed for the support containing 15% porosity. The catalysts occupied the surface, the microprobes and the macropores of the supports, when the catalyst loadings were close to the porosities of the supports. The surface, the micropores and macropores of support were mostly filled up by the catalysts, which resulted in decreases in the macropore and the micropore volumes. However, a high level of large macropores(diameter more than 0.1mm)facilitates the intraparticle diffusion, and micropores(diameter less than 20nm) are necessary to develop a high surface area. Hence the conversion decreased slightly, after it reached the highest point.

THE EFFECT OF DIFFERENT CONSTITUENTS OF CATALYSTS- Fig.2 shows that the conversion of CO and C_3H_8 were different with the active constituents at $25000h^{-1}$ and $5000h^{-1}$ respectively. The addition of the fourth active constituent had effects. In the CeCuMn and CeCuZr system, combing the Zr and the Mn had little effect on the conversion, but with CeCuM'M" the conversion increased significantly. The fourth constituent increased the "active site", which may be viewed as the point on the catalyst material crystallite where the electronic forces are optimum for the catalytic reaction to take place. The catalytic process is illustrated by the following example: CO and O_2 were chemisorbed on the catalyst and can react readily because of their proximity and orientation[3,4]. The process of adsorption also results in a weakening of the bond between the atoms within the CO molecule because some of the energy is shared with the surface. Thus, the adsorbed atoms of the molecule were less tightly bonded to the molecule and more easily attracted to other atoms such as oxygen. The reaction between CO and O_2 was thus easier and more rapid.

The basic requirement of such a catalyst is that it chemisorbes the molecules in the desired temperature range and in such a way that the reaction occurs readily. Then, the products having achieved a lower energy state must desorb at the same temperature, freeing the active site for additional reaction.

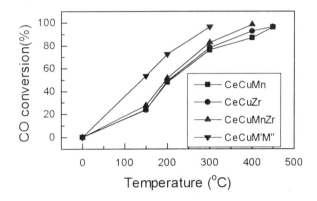

Fig.2 Effect of different components of catalyst

Table1 shows the radial distribution of the rare earth Ce, transition metals Cu, M' and M". The Ce, M' and M" were basically homogeneous throughout the beads, but Cu increase gradually from the centre to the outside of the pellets.

Table 1. The radial distribution of elements

mm	Al	Ce	Cu	M'	M"
0.2	53.40	34.45	1.07	7.39	2.69
0.5	53.97	33.86	2.30	6.08	3.80
1.0	54.02	34.82	3.91	5.08	2.16
1.5	53.18	32.16	6.19	5.84	2.63

THE EFFECTS OF PELLETS SIZE-The design of the pellets must include the necessity of achieving high efficiency at very low residence times. Consequently the lowest resistance to mass transfer from the gas flow to the catalytic material surface is required. Fig. 4 shows that small pellets were more efficient than the bigger ones at low temperatures. This is attributed to the small radius pellets providing a large contact area between the beads and gas, and reducing the intraparticle distance to the active site, which determined the conversion at low temperature. At the higher temperatures, the conversion mainly depended on the activity of the catalysts , so the conversion was almost equal.

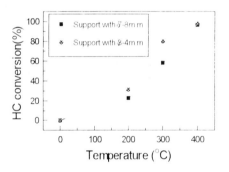

Fig.4 Effect of the size of pellets

SUMMARY

When the loading amount of catalysts were close to the porosity of the support, the micropores and the macropores were almost filled up by the catalysts, and the conversion reached the highest point. The conversion then decreased slightly with further increases in the catalyst loading.

The fourth constituent increased the "active site". Over the CeCuM'M" catalyst at $25000h^{-1}$, the CO could be oxidized about 53.4% at 150°C. The radial distribution of the rare earth Ce, transition metals M' and M" were homogeneous, but Cu increased gradually from the centre to the outside of the pellets.

Small pellets were more efficient than the larger ones at low temperatures. This is attributed to the small radius pellets providing a large contact area between the beads and gas, and reducing the interparticle distance to the active site, which determined the conversion at low temperatures.

REFERENCE

1. Caroled L A et al., SAE Technical Paper No.892040,1989.
2. Hiroaki Yamamoto et al., SAE Technical Paper No.900614,1990.
3. Daniel Durand et al. US. Pat.5108978,1992.
4. Yoshiasu Fujitani et al.,US Pat.4367162,1983.
5. Shyu J Z,Otto K, J.Catalysis,1989;115:16
6. Shin-ike T., Saki T.,Mat.Res.Bull.,1977;12:269
7. Mo Carthy T.,Kenneth E, Mat.Res.Bull.,1974;9:1279
8. Obayashi H.,Kudo T.,Mat.Res.Bull.,1978;13:1409
9. Qin shunchang et al.,Proceedings'94 Baotou International Symposium on Rare-Earth Science and Technology,1994

961133

Changes in Pollutant Emissions from Passenger Cars Under Cold Start Conditions

Robert Joumard and Robert Vidon
Institut National de Recherche sur les Transports et leur Sécurité

Laurent Paturel
Savoy Univ.

Gérard de Soete
Institut Français du Pétrole

Copyright 1996 Society of Automotive Engineers, Inc.

ABSTRACT

CO, CO_2, HC, CH_4, NOx, N_2O emissions as well as emissions of 10 PAHs and fuel consumption were measured on 10 conventional petrol-engined passenger cars, 10 vehicles equipped with 3-way catalysts, and 5 diesel-engined vehicles over a great number of driving cycles under hot or cold start conditions : the 2 standardized European and American cycles, Highway cycle, 4 hot cycles representative of real-world driving conditions, and 3 representative mini-cycles, which have been repeated 15 times after a cold start. Simultaneously, water and oil temperatures were measured to assess engine temperatures. The analysis of the results enabled the monitoring of emission and temperature changes with time for various kinematic types. The limit of hot temperature, the distance travelled under cold start conditions (over about 6 km), the excess emission rates under cold start conditions, and absolute excess emissions after a cold start could be thus determined. These parameters significantly vary according to the vehicle technology, driving cycle and pollutant type. In addition, a great number of results are presented relative to hot emissions and standard cycles.

INTRODUCTION

Pollutant emissions are admittedly significantly higher with a cold engine, i.e when the engine has not reached its equilibrium temperature, in particular with regard to petrol engines, or when exhaust gases have not reached a minimum temperature for good operating conditions of the catalyst [1][2]. Excess emissions are generally assessed by comparing hot and cold start emissions over a same, often standardized, cycle.

As about 30% of the mileage is travelled with a cold engine [3], the quality of such an evaluation is very important. When the driving cycle selected is too short, only very cold emissions with abusively high excess emission rates are taken into account. Conversely, if the cycle is too long, excess emission rates recorded are abusively too low. Thus, the authors wished to determine with accuracy the temperature changes, and more particularly the emission changes, under usual ambient temperature conditions, for three types of passenger cars (petrol, catalyst and diesel vehicles).

This research study enabled us to improve available knowledge respective to actual unit emissions of the European in-operation fleet considering its various components. This is the continuation of a previous research [1] focussed on the impact of vehicle parameters and average speed on emissions. Finally, the last objective is to test the representativity of standardized cycles in Europe and the United States relative to cold and hot emissions. A detailed report relating to this research is available [4].

EXPERIMENTAL PROCEDURE

VEHICLE SAMPLE - 25 private sample cars (10 petrol-engined cars not equipped with catalysts, 10 with a 3-way catalyst complying with US83 standard and 5 IDI diesel-engined vehicles, including 2 turbo charged) were tested. The three sub-groups sampled meet the following parameter distribution in the 1995 French fleet: emission standard, engine capacity and make (some vehicle characteristics are listed in Table 1). In addition to these three criteria the vehicles were selected by random sampling among vehicles proposed by theirs owners. The vehicles selected were mainly privately-owned vehicles (21) and four of them were hired vehicles. Petrol and diesel engined vehicles were selected from the French fleet, catalyst vehicles from the Swiss fleet. Tests were performed on as-received vehicles. The fuels used were bought in local refuelling stations.

DRIVING CYCLES - Vehicles were tested on a chassis dynamometer. Vehicle cooling, and more particularly engine cooling, is provided by a fan and

fuel catalyst		petrol without	petrol with	diesel without
number		10	10	5
cubic capacity	(cm³)	1386	1674	1755
power	(kW)	50.6	66.5	49.7
mass	(kg)	873	978	986
mileage	(km)	44 742	38 335	59 606

Table 1: Average values of some technical characteristics of test vehicles.

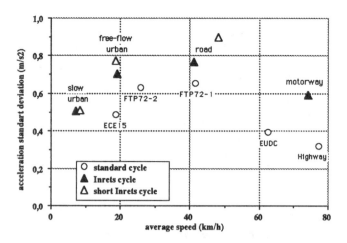

Figure 1: Driving cycle description as a function of average speed and acceleration standard deviation.

ambient air flow is regulated as a function of the vehicle speed, over an area of 0.8 m². Cooling conditions are thus close to those observed under on-the-road driving conditions.

For each vehicle four types of driving cycles were followed:

- standardized cycles performed under cold start conditions: European NEDC cycle (ECE 15 followed by the EUDC cycle), and the FTP75 cycle (FTP72-1 followed by FTP72-2 and then FTP72-1). These cycles were followed in accord with standard requirements: ECE15 is started 40 s after vehicle starting and the vehicle is shut off for 600 s between the second and the third sub-cycles for FTP75;

- a standardized cycle performed under hot start conditions: the US Highway cycle;

- 4 cycles representative of real use conditions of the vehicles in France, called Inrets cycles, under hot start conditions: slow urban, free-flow urban, road and motorway, with resp. average speed of 7.3, 19.2, 41.2 and 74.2 km/h [1][5]. Duration of such cycles is 12 to 18 minutes;

- 3 short cycles - short slow urban, short free-flow urban and short road - defined from corresponding Inrets cycles and provided with same kinematic features. Duration of short Inrets cycles is 2 to 3 minutes. Each of these short cycles was repeated 15 times, the first of them being started with a cold engine.

Gear ratios were imposed for standardized cycles, and calculated for each vehicle as related to Inrets cycles and Inrets short cycles. In the latter case, gear ratio distribution as a function of speed and acceleration parameters correspond to the distribution observed during instrumented surveys over fleet vehicles.

Fig. 1 describes the cycles versus average speed and acceleration standard deviation. For standardized cycles the accelerations recorded are significantly lower than those of Inrets cycles but short cycles recordings are relatively close to those of representative Inrets cycles.

The laboratory is not provided with a room thermostat and ambient air temperature recorded in the laboratory is slightly higher than outside air temperature. Laboratory average temperature is 15.6 ± 3.9°, i.e 13.0, 16.1 and 19.6° C for petrol, catalyst and diesel engined vehicles respectively. Temperature variations between these three engine types were liable to have an impact on comparative analysis results. But from a previous study [1], ambient temperature has a substantial impact on emissions (not including NOx) from diesel vehicles only. Air pressure amounts to 994 hPa and humidity to 50%. Under cold start conditions, vehicles are at laboratory temperature.

MEASURED PARAMETERS - Cooling water and oil temperatures were continuously measured, and for catalyst vehicles the temperature of the exhaust pipe surface located before the catalyst, usually called exhaust gas temperature, was also recorded.

Exhaust gases were sampled using a CVS, by continuous sampling and using bags or filters. Pollutants were analyzed using conventional methods (infrared absorption for CO, CO_2, and CH_4; flame ionization for total HC, luminescent chemistry for NOx, weighing for particulates); by calculation (fuel consumption) or by specific methods (N_2O; polycyclic aromatic hydrocarbons - PAH). Methane emissions of catalyst vehicles have not been analysed. It should be observed that PAH have not been recorded for all cycles (only for free-flow urban, road, short free-flow and short road cycles) and have been the subject of a unique sampling for all 15 short cycles, while particulates have been analyzed only for diesel vehicles over long cycles (standardized and representative ones).

N_2O was analyzed by gas chromatography with electron detector [6], after sampling bag transfer to the IFP laboratory.

PAH sampling was performed in the CVS dilution tunnel using a filtering (Teflon wool) and absorbing (XAD2 Amberlite resin) device, well adapted to collect particulate and gaseous PAHs [7]. They were then transfered to the Savoy University laboratory.

All PAHs are extracted by cyclohexane ultra-sound method. After dry evaporation, samples are taken into the octane which is the Shpol'skii matrix used. PAH identification and sampling were performed by high resolution spectrofluorometry, at 10 K temperature, using specific equipment and method [8][9].

13 PAHs were analyzed: 3 including 4 nuclei, pyrene (P), benzo (a) anthracene (BaA) and chrysene (Chr); 8 including 5 nuclei, benzo (b) fluoranthene (BbF), benzo (j) fluoranthene (BjF) and benzo (k) fluoranthene (BkF), benzo (a) pyrene (BaP), perylene (Per), dibenzo (a, h) anthracene (DBA), benzo (e) pyrene (BeP) and benzo (b) chrysene (BbC); 2 including 6 nuclei, indeno (1,2,3 - cd) pyrene (IP) and benzo (ghi) perylene (BghiP).

Chr, BeP and BbC PAHs have never been detected, while DBA, Per, IP, BbF and BjF PAH have rarely been detected for catalyst petrol vehicles over the 2 hot cycles because of the low quantity of PAHs sampled. In the three cases studied, the 5 BkF, P, BaA, BaP and BghiP have always been quantified or detected. Thus, 3 PAHs have never been detected, 10 have generally been detected and 5 have systematically been detected. This document gives emission values for 10 PAHs and their summation noted 10HAP and the summation of the 5 PAHs always detected, noted 5HAP.

All emission and consumption results are given in mass unit per travelled distance unit (g/km or µg/km for PAH). CO, CO_2, CH_4, N_2O, PAH, fuel consumption and particulates are expressed in real mass emitted, while HC and NOx are expressed in equivalent mass of CH_4 and NO_2 respectively. In this report instantaneous emissions values (mesured continuously) are not studied and only emissions measured over a driving cycle or over a set of driving cycles are reviewed. Finally only average results are considered for each of the three types of vehicles: the low number of vehicles tested per sample does not allow a refined analysis of the results.

RESULTS : CONVENTIONAL EMISSION FACTORS

Only conventional standardized cycles (NEDC, FTP75, Highway) or Inrets cycles (4 cycles) are considered. Each cycle is considered as a cold cycle if started under engine cold conditions (ECE15 and first FTP72 1 cycles), and as a hot cycle in all the other cases (EUDC, FTP72-2, second FTP72-1, Highway, 4 Inrets cycles).

EMISSION AS A FUNCTION OF THE CYCLE - Figure 2 shows an example of the results obtained versus cycle average speed. Conventional impact of average speed, differences between standardized cycles under low acceleration conditions and the most representative Inrets cycles in terms of driving behaviours are given, as well as a first indication about cold start impact. But the lack of emission measurements over a same cycle started with a hot and a cold engine does not enable an accurate evaluation of cold start impact, even for FTP72-1 cold and "hot", because it is not sure that the last one is hot.

AVERAGE HOT EMISSIONS - In order to compare the three vehicle types in terms of emission levels, it is proposed to sum up some of the results recorded and to weigh them. All Inrets cycles correspond to vehicle actual usage conditions, each cycle corresponding to a prescribed portion of total mileage [5]. Hot emissions values measured were weighed for the four Inrets cycles, using 0.01, 0.16, 0.42 and 0.41 factors respectively, for all the pollutants, excepting PAHs. With regard to PAHs, only Inrets free-flowing and road cycles were tested. Weighing factors are 0.17 and 0.83 respectively, corresponding to the above urban and extra-urban weighting factors.

Average unit emissions representative of all usage conditions are then obtained. The results recorded are given in Fig. 3 for all the pollutants measured. But comparative results given for the three technologies presented below are limited due to the low number of sampled vehicles.

CO, HC and NOx emissions of catalyst vehicles are close to those emitted by diesel engined vehicles; but CO and HC emissions are 20 times lower and NOx emissions are approximately 4 times lower than for petrol vehicles not equipped with catalysts, while CO_2 emissions are increased by 22%, due to the oxidation of the CO and HC in the catalyst and maybe the increasing vehicle weight, which affects the fuel consumption. The most significant results concern methane, N_2O and PAH emissions, for which there is a very few data in the literature. CH_4 emissions for diesel vehicles are significantly lower than for petrol vehicles (by about a factor of 9), and N_2O emissions are reduced by about 25% only. On the other hand, N_2O emissions of catalyst vehicles are twice as high as those of non-catalyst vehicles.

As regards PAHs diesel vehicle emissions are

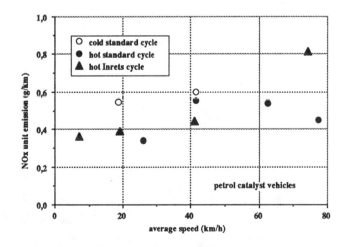

Figure 2: Average NOx unit emissions for catalyst petrol vehicles versus average speed for conventional driving cycles (standardized and Inrets). Refer to Fig. 1 to identify the various cycles.

significantly lower than those of non-catalyst petrol vehicles, and significantly higher than those of catalyst vehicles. The variations observed for non-catalyst petrol vehicle and diesel vehicle emissions are close to those described previously [1][7], with an average emission value about twice as low for diesel vehicles. These values are higher at high speeds, by about 20 to 35% as a function of the cycle studied, with a speed impact which is less significant for diesel vehicles.

Simultanously the values recorded for 5 PAHs for catalyst petrol vehicles are very low: only 6.7% of those of non-catalyst vehicles (with a higher percentage at low speed: 12.1 and 5.9% respectively for hot free-flow urban and road cycles). This result can be compared to the values reported by a number of authors [10].

Nevertheless technology variations differ significantly as a function of PAH studied, without modifying substantially the measurement scale. Thus, P, BkF and mainly BaA are less reduced than other PAHs, catalyst action being more effective for heavier hydrocarbons. For diesel powered vehicles, as compared to petrol engined vehicles, the emissions are about the same for BkF and BbF, while diesels emit less of the other PAH's than petrol vehicles. This trend is the result of an engine temperature increase, thus favouring cycling and higher condensation hydrocarbon emissions for petrol engined vehicles.

PARTICULATE AND GASEOUS PAH DISTRIBUTION - Eight analyses (2, 2 and 4 analyses for petrol catalyst, non-catalyst vehicles, and diesel vehicles respectively) were carried out on samplings performed over the whole 15 successive short road cycles after a cold start in order to determine particulate and gas PAH distribution, for each vehicle type. The ten PAHs were systematically detected for the two sampling phases.

PAHs obtained in the particulate phase for petrol catalyst and non-catalyst vehicles account for a very significant part of the overall PAH mass, which is substantially equivalent for the two technologies studied, of about 70 and 65% respectively. Thus, the particulate phase is the most significant, even for petrol engined vehicles.

The amounts obtained in the particulate phase for diesel vehicles are even more significant, ranging from 80 to 90%, for an average value of about 82.5%. Percentage changes in volatile PAH emissions seem to be connected to the increase in hydrocarbon molecular weight.

ENGINE TEMPERATURE IMPACT ON EMISSIONS

Engine temperatures can be considered alone, i.e maximum temperature values that could be used to define engine steady operating conditions. Nevertheless, it is not certain that emission stability corresponds to temperature stability: from an emission standpoint, a "hot" engine has been defined as an engine under steady emission conditions.

The study of temperature changes as functions of time and travelled distance (see an example Fig. 4) demonstrates that:
- cooling water reaches equilibrium temperature quicker than oil;
- equilibrium temperatures are not dependent upon the driving cycle, but differ significantly for water and oil according to the vehicle types;

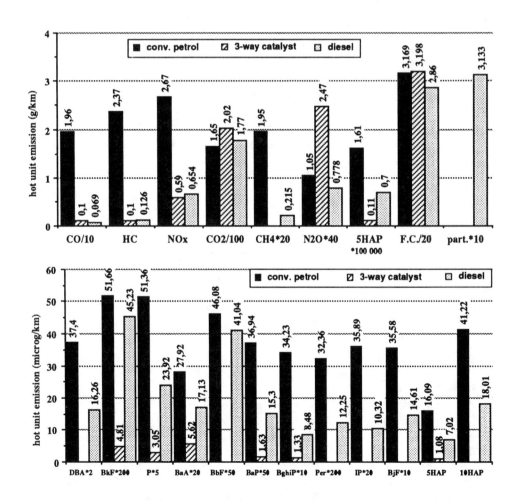

Figure 3: Actual hot average unit emissions according to vehicle category.

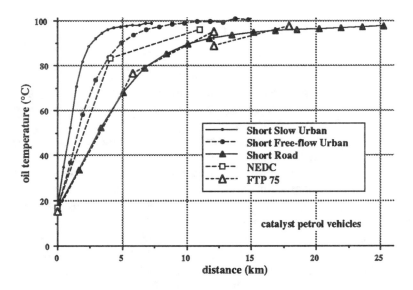

Figure 4: Changes in engine oil temperature for petrol-fueled catalyst vehicles, versus distance.

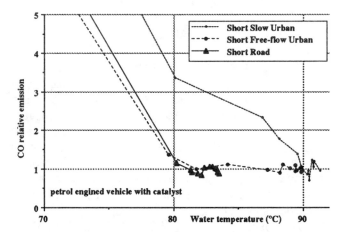

Figure 5: Evolution of relative CO emissions versus cooling water temperature for vehicles equipped with a catalyst.

- the driving cycle has an impact on trip duration and travelled distance at a temperature below the equilibrium temperature. Duration is nevertheless a less discriminating factor respective to the cycle studied than distance. As unit emissions are expressed per km travelled, priority was given to cold start analysis as a function of travelled distance, even if the distance travelled with a cold engine is a highly varying factor;
- standardized NEDC and FTP75 cycles are in good agreement with short Inrets cycles in terms of temperature.

From a pollution standpoint, engine temperature recording is not a very interesting parameter. The problem is to know whether pollutant emissions have reached their equilibrium temperature level (hot emissions) or not (cold emissions). The relationship between engine temperature and pollutant emissions must be thus considered. Curves of emission changes versus cooling water temperature (an example is given in Fig. 5) and oil temperature demonstrate that changes differ according to studied pollutant, vehicle types and driving cycles.

Mimimum temperatures, from which emissions are considered steady, can be thus deduced. Table 2 gives water, oil and exhaust gas temperatures per pollutant and vehicle type. This includes fuel consumption parameters and all the pollutants, excepting PAH and particulate matters. Indeed, the sampling type selected for PAHs does not enable an analysis with time, while particulates have not been measured during short cold start cycles. These temperatures are dependent upon the driving style: therefore limit temperature ranges are given. Globally, limit temperatures are as follows:
- for water:
    - close to 88° for petrol engined vehicles, except for total hydrocarbons,
    - close to 86° for catalyst vehicles,
    - close to 70° for diesel engined vehicles,
- for oil:

	Vehicle	CO	HC	NOx	CO_2	CH_4	N_2O	F.C.	Σ
Water	Petrol	87-90	81-87	87-89	85-90	87-89	-	85-87	81-90
	Catalyst	82-90	82-89	82-90	83-90		82-91	82-90	82-91
	Diesel	67-75	67-73	67-73	67-72	68-75	-	67-74	67-75
Oil	Petrol	85	60-90	80	80-90	90	-	80-90	60-90
	Catalyst	93	75-95	92-95	93-95		88-97	92-97	75-97
	Diesel	88-96	92-98	67-98	83-90	82-100	-	60-85	67-100
Exhaust gas	Catalyst	225	220-227	225-240	225-240		227-235	225-240	220-240

Table 2: Water, oil and exhaust gas limit temperatures beyond which vehicle emissions are steady, in °C.

- close to 85° for petrol engined vehicles,
- close to 92° for catalyst vehicles,
- close to 90° for diesel engined vehicles,
- for exhaust gases (pipe surface):
  - close to 230° for catalysts. Bed temperatures in catalyst are probably much higher.

Nevertheless it should be noted that these limits are only approximate values and actual recordings may significantly vary. Water temperature seems to be a better indicator of engine thermal condition in terms of emissions since, for each technology studied, limit temperature variations are lower for water temperature than for oil temperature. Limit oil temperatures significantly vary as a function of the driving cycle and the pollutant studied.

Temperature estimates below which cold emissions are recorded enable to confirm or not the "cold" or "hot" character of any emission, considering the kinematic type. This is specifically the case with non initial portions of standardized NEDC and FTP75 cycles after a cold start. EUDC cycles in the first case, and FTP71-2 and terminal FTP71-1 in the second case are generally considered hot from an emission standpoint. Is that really the case?

Table 3 gives initial and final temperatures for each of the three sub-cycles studied. When comparing these temperatures with hot temperatures measured over short free-flow urban (18.8 km/h) and short road (48.1 km/h) cycles, respectively for FTP72-2 on one hand, and EUDC and FTP72-1 on the other with close average speeds, it can be observed that none of the three standardized cycles is systematically cold or systematically hot for all the pollutants studied. If considering in priority the most sensitive pollutant to engine temperature (CO, HC and, for catalysts only, NOx), it can be seen that:

- the EUDC cycle and the FTP72-1 cycle (last portion of the FTP75 cycle) are generally started under hot conditions respective to cooling water and cold conditions respective to oil, and are generally ended under hot conditions,
- the FTP72-2 cycle is started most often under cold conditions, and ended under hot conditions for petrol engined vehicles equipped or not with catalysts, under cold conditions for diesel engined vehicles.

It can be therefore asserted that the first portions of NEDC and FTP75 cycles are undoubtedly cold, but do not cover the whole cold period due to cold start conditions. Thus subsequent cycles cannot be considered "hot" since they are performed partly over the cold period. It corresponds to the conclusion of Lenner [11], who compared FTP75 with and without 600 s pause. Each standardized NEDC and FTP75 cycles are nevertheless ended under hot conditions and are sufficient to warm up light vehicles.

These examples demonstrate that usual cycles, whose duration is close to that of standardized ECE15, EUDC, FTP72-1 or FTP72-2 cycles, do not enable a correct warming-up of the vehicle and do not fully cover the cold period after a cold start.

EMISSION CHANGES AFTER A COLD START

To study accurately emission changes after a cold start, it is required to consider as far as possible speed variations, since, as already seen, hot temperatures are highly dependent upon the driving cycle. Therefore, very short driving cycles were developed under similar speed and cold start conditions and were repeated up to reaching steady emission values. These are short Inrets cycles which have been repeated 15 times after a cold start. Each short cycle is characterized by a prescribed emission level for each studied pollutant.

VARIOUS TYPES OF EMISSION CHANGES - From cold start, some changes in emission levels are recorded (continuous decrease or even increase). These levels

		EUDC begin	end	"hot" FTP72-2 begin	end	"hot" FTP72-1 begin	end
water temperature	Petrol	85	<u>89</u>	82	<u>89</u>	82	<u>89</u>
	Catalyst	<u>87</u>	<u>87</u>	86	<u>91</u>	<u>87</u>	<u>88</u>
	Diesel	<u>67</u>	<u>66</u>	***66***	70	<u>69</u>	<u>71</u>
oil température	Petrol	75	<u>90</u>	70	<u>90</u>	83	<u>93</u>
	Catalyst	83	96	***77***	<u>95</u>	89	<u>98</u>
	Diesel	80	95	***81***	95	86	<u>99</u>
exhaust temp.	Catalyst	***179***	<u>245</u>	***219***	226	***134***	224

Table 3: Temperatures at the beginning and the end of "hot" sub-cycles of European and US-standardized cycles (°C). **Bold italic** characters : systematically cold temperatures; <u>underlined</u> characters : systematically hot temperatures whichever the pollutant.

are then stabilized around the hot value, with some variations, cycle after cycle. The end of the cold period can be assessed by calculating emission standard deviation from the last cycle period up to the initial one. In the case of nearly steady hot emission value, standard deviation decreases as the number of measurement points, considered back in time, increases. When reaching the cold period, a clear emission variation which is not a random value is recorded. Then the standard deviation continuously increases when going back in time.

This technique enabled us to assess the number of short cycles performed under cold conditions and then the number of short cycles actually performed under hot conditions. This yields an average hot emission value, and then for each short cycle a relative emission value, divided by the hot emission value. Only this emission value will be considered here for an easier comparative analysis of the pollutants.

The curve shapes of emission changes can be easily differentiated as functions of the pollutants and vehicles studied. Examples are given in Fig. 6:

- a continuous decrease is the most frequent change recorded: this is systematically observed for HC, CO_2 and fuel consumption; this is also the case of catalyst vehicles for CO, and diesel vehicles for CO and CH_4: this is a conventional excess cold emission value;
- a decrease followed by a slight increase to reach the hot value: this is the case of petrol engined vehicles for CO and CH_4, of catalyst vehicles for NOx and N_2O. Thus an excess emission followed by a reduced emission are noted;
- an increase followed by a slight decrease in NOx emissions for petrol engined vehicles: a reduced emission followed by an excess emission are noted;
- in a number of other cases, which are very few (NOx and N_2O for diesel engined vehicles), a continuous emission increase can be observed, but it is not very significant nor clear.

Therefore it seems that it is not correct to speak systematically of an excess cold emission, even if it is the most frequent case.

DISTANCES TRAVELLED UNDER COLD CONDITIONS - A cold emission can be characterized by the change curve evolution, as already seen, as well as by its duration or travelled distance, and eventually by the relative emission rate over the whole period.

Table 4 gives the distances travelled under cold conditions: these vary significantly, ranging from 0 to 19 km. As for durations, cold distances depend on the driving cycle, vehicle technology and the pollutant considered. Unit emissions are usually expressed per kilometer travelled and we mainly considered the distance covered under cold conditions.

With regard to the driving cycle, the distance is nearly systematically an increasing function of cycle average speed, while cold duration is a decreasing function of it. If the results obtained are averaged over all the pollutants considered, the relationship appears clearly.

To discard speed influence, distances travelled were averaged under cold start conditions over the three short cycles : specific cold distance values were obtained for each pollutant and each vehicle technology, as shown in Fig. 7. Cold distances vary for regulated pollutants (CO, HC, NOx) from 3.2 to 8.2 km with an average value of 5.7 km; but these can reach 12 km for CH_4 and N_2O.

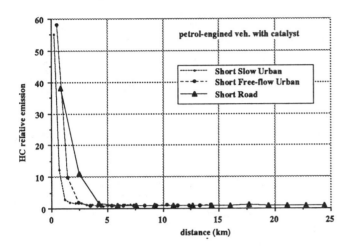
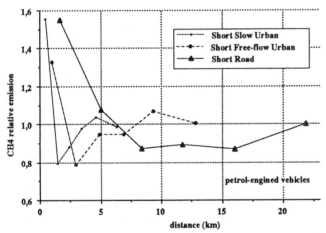
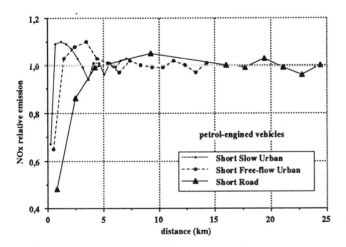

Figure 6: Typical emission changes versus travelled distance after cold start.

	speed (km/h)	CO	HC	NOx	CO_2	CH_4	N_2O	F.C.	average
Petrol	8.4	4.4	3.9	2.4	4.4	2.9	5.3	1.9	3.6
	18.8	7.9	2.0	3.9	4.9	3.9	0	6.9	4.2
	48.1	6.7	6.7	3.4	10.1	18.5	13.5	13.5	10.4
Catalyst	8.4	2.9	2.9	2.9	1.9		5.3	1.9	3.0
	18.8	3.0	3.0	9.9	5.9		3.9	5.9	5.3
	48.1	11.8	5.1	11.8	13.5		18.5	11.8	12.1
Diesel	8.4	4.4	4.4	4.9	2.4	5.3	5.3	5.3	4.6
	18.8	7.9	7.9	8.9	3.9	10.8	10.8	4.9	7.9
	48.1	8.4	8.4	3.4	11.8	18.5	18.5	10.1	11.3

Table 4: Distances travelled under cold conditions (in km) according to pollutant type, vehicle type and average cycle speed (in km/h).

To assess the influence of vehicle technology, these last values are integrated over all the pollutants involved or over regulated pollutants only: the results show that, in all the cases studied, the lowest cold distances were recorded for petrol engined vehicles and highest ones for diesel engined vehicles, catalyst vehicles showing intermediate distances covered under cold conditions. On average it can be generally considered that after a cold start, emissions do not reach a steady value before 6 to 7 km. But for regulated pollutants, this cold distance is slightly shorter: it is close to 4.5 km for petrol engined vehicles, to 6 km for catalyst vehicles and close to 6.5 km for diesel vehicles.

EXCESS EMISSION LEVELS - In the previous section, we assessed the cold distance travelled, i.e from vehicle start-up to emission stabilization. Then, an average cold unit emission was obtained over the same distance, followed by a hot unit emission beyond the cold period. This yields a relative cold excess emission value (cold unit emission / hot unit emission ratio), and an absolute cold excess emission, i.e respective to the pollutant mass specifically due to cold start, which is the additional emission value obtained under cold conditions compared to the emission value that could have been recorded for the same period under hot conditions.

Excess emission values vary significantly from one cycle to another, from one technology to another, and of course according to the pollutant considered. To suppress speed influence, average relative emission value recorded during the cold period is calculated over the three cycles monitored and weighed by cold distance: see Fig.7. The results obtained show that:

- excess emissions for vehicles equipped with catalysts are very significant for CO and HC, by a factor close to 10 and 16 respectively, over distances of 5.9 and 3.6 km respectively;
- for petrol and diesel engined vehicles, low NOx emission decreases (- 13% and - 4% respectively for distances close to 3 and 6 km); and N_2O emission decreases (-5 and -2% over 6.3 and 11.6 km respectively);
- relatively significant excess emissions in all other cases, ranging from 4 to 88% over distances ranging from 4 to 12 km.

Figure 7 gives simultaneously average absolute excess emissions, which are the average of absolute excess emissions recorded over the three cycles studied and weighed by the representativity factor of each cycle in the overall mileage covered by the vehicle fleet. The results are analyzed in tems of technology later.

PAH were not sampled for each short cycle. Therefore, this cannot yield the cold distance value. Two assumptions were thus made:

- with regard to PAHs, the distance covered under cold conditions is similar to that covered under cold conditions for total hydrocarbons HC;
- hot unit emission over short Inrets cycles is equivalent to the unit emission recorded over conventional hot Inrets cycles, despite substantial differences in kinematic quality.

These assumptions enabled us to calculate a cold unit emission for each of the ten PAHs, and then a relative cold excess emission and an absolute cold excess emission value, as performed for the other pollutants, per vehicle type and for the two cycles followed.

Figure 8 gives relative and absolute average excess emission values (weighed by the cold distance value) for each PAH, for 10HAP and 5HAP summations and per vehicle type.

For vehicles equipped with catalysts, only average relative cold emission and average absolute excess emissions of the 5 PAHs and the 5HAP summation can be considered significant; since the other PAHs have not been systematically detected over reference hot cycles. For the whole 5 PAHs, cold relative emission is a very important parameter, since it varies from 2.6 to 5.4 as a function of the technology studied. These relative excess

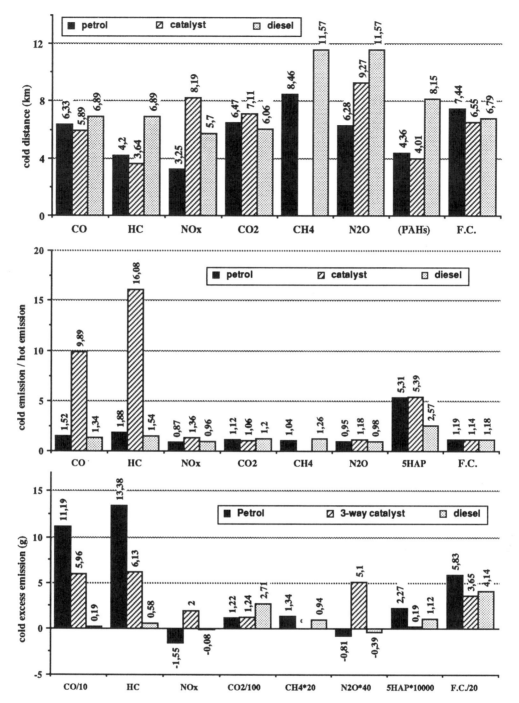

Figure 7: Distance travelled under cold conditions (top), cold / hot unit emission ratios for the whole cold period (centre), and absolute cold excess emission (bottom), according to pollutant and vehicle types (last written digits not significant; thus for petrol engined vehicles, CO unit emission over a cold period of 6.33 km is 52% higher than hot unit emission, and yields an excess emission of 11.2 g).

emission values are higher for catalyst and non catalyst petrol vehicles, with very high values at lower speeds by a factor of about 14 and 7 respectively. For diesel vehicles, this is even more noticeable at high speeds with an average weighed value amounting to 2.6. Corresponding absolute excess emission values for the 5 PAHs are of about 230, 20 and 110 µg respectively for petrol, catalyst and diesel vehicles, hot unit emissions being very low for catalyst vehicles. As for hot emissions, absolute cold excess emissions of the 5 PAHs for diesel vehicles are intermediate values between these of catalyst and non catalyst petrol engined vehicles.

COMPARING PETROL-CATALYST-DIESEL VEHICLES - Absolute excess emissions given in Fig. 7 for all the pollutant studied and in Fig. 8 for PAHs were used to compare the three technologies studied. Results

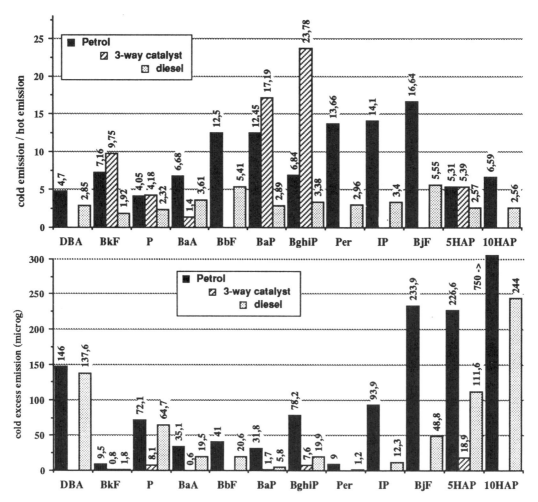

Figure 8: Relative cold emissions, and weighed cold absolute excess emissions for the various PAHs studied and their summation.

vary significantly as a function of the pollutant analyzed. Nevertheless the following conclusions - to be relativized owing to the small number of sample vehicles and in particular diesel vehicles - can be drawn:

- diesel vehicles generate the lowest cold excess emissions, except for CO_2 and the various PAHs. For most pollutants, diesel vehicles are less sensitive to engine thermal condition than petrol vehicles;

- for catalyst vehicles, despite relative emissions sometimes very high, absolute excess emissions are relatively low, located between the values of the two other technologies studied, except for NOx and N_2O for which excess emissions are the highest and PAHs for which excess emissions are systematically lower;

- non catalyst petrol vehicles are more sensitive to cold start in terms of CO, HC, CH_4, PAH emissions and fuel consumption.

Comparing distances covered under cold conditions, relative emissions and absolute cold excess emissions remains a partial analysis. To assess overall performance of the various technologies in terms of cold emissions, cold emissions themselves must be compared. As the distances travelled with a cold engine differ from one technology to another, it should not be correct to compare absolute emissions in grams during the cold period.

For an accurate comparison, average unit emissions were calculated from start-up for each short cycle sequence. This enabled to compare pollutant emissions for any trip started with a cold engine and to highlight the polluting contribution of a given vehicle technology or the other to perform a trip with a cold engine. It has been therefore demonstrated that:

• *For CO and HC:*

- non catalyst petrol vehicles generate significanlty higher emissions that catalyst or diesel vehicles. Even in the first hundreds metres, emissions of a catalyst vehicle are twice as low as that of a non-catalyst vehicle;

- for diesel vehicles, CO emissions are always lower. HC emissions are nearly always lower than those of catalyst vehicles. Distances amounting to 36, 70 or 114 km (for short slow urban, short free-flow urban and short road cycles respectively) should be covered to obtain cumulated HC emissions from diesel vehicles exceeding those of catalyst vehicles;

• *For NOx:*

- for catalyst vehicles emissions are higher than those of non catalyst petrol vehicles only over very short distances after a cold start: 0.8, 1 and 2 km respectively for the three short slow urban, free-flow and road cycles;

- for diesel vehicles, emissions are generally higher than those of catalyst vehicles, except over the first kilometers at high speed (over 3 and 14 km for short free-flow urban and road cycles) respectively;

• *For CO_2:*

- catalyst vehicles are nearly always more polluting; only diesel vehicles generate high emissions over the first 14 kilometers of the short free-flow urban cycle and over the 3 first kilometers of the short road cycle. The vehicles not equipped with catalysts are always less polluting than catalyst vehicles, primarily due to the higher HC and CO emissions;

- diesel vehicles release more CO_2 than non catalyst petrol vehicles, except at very low speeds (case of the short slow urban cycle); this is not dependent upon engine thermal condition;

• *For CH_4:*

- emissions of petrol vehicles are systematically 10 times higher than those of diesel vehicles, under cold or hot conditions;

• *For N_2O:*

- in all the cases, hot and cold emissions of catalyst vehicles are twice or three times higher than those of conventional petrol or diesel vehicles. Diesel vehicles always generate lower emissions than petrol vehicles, except at very low speeds (case of the short slow urban cycle) for which emissions are equivalent. Engine thermal condition is not a significant factor for this comparative analysis;

• *For fuel consumption:*

- fuel consumption (mass value) of diesel vehicles is always the lowest when comparing the three technologies studied, except over the first kilometers of the short free-flow urban cycle (first kilometer as compared to petrol vehicles, the first five kilometers as compared to catalyst vehicles);

- fuel consumption of catalyst vehicles remains lower than that of non-catalyst vehicles over the first 7 km at very low speed (8.4 km/h), over the first 18 km of the road cycle, and systematically lower for the free-flow urban cycle, under hot and cold conditions.

It can therefore be asserted that, although cold start has a contrasted influence on the three technologies studied, their classification is not significantly modified in terms of emissions, as usually established for hot conditions.

The first noticeable exception relates to oxides of nitrogen NOx, for which higher emissions are recorded with catalyst vehicles as compared to conventional vehicles over the first kilometer. This drawback of catalyst vehicles is widely compensated by their performance beyond the first kilometer.

The second noticeable exception concerns fuel consumption for which the petrol / catalyst vehicle comparison yields highly varying results as a function of the vehicle speed and engine thermal condition. Averages recorded depend mainly on weighing assumptions made between cold and hot engine operating conditions, the discrepancy always remaining very low.

SUMMARY AND CONCLUSIONS

1 - Repeating the short driving cycles yielded engine temperatures and emission for the various pollutants measured after a cold start at ambient temperature and under realistic speed conditions, representative of real-world driving conditions.

2 - Water or oil temperatures, which correspond to steady emissions, thus defining "hot" emissions, vary significantly as functions of used technologies, driving style and even the considered pollutant. It is therefore difficult to determine whether the engine is cold or hot from a pollutant emission standpoint from engine temperatures only.

3 - This allowed us to define the meaning of "cold" and "hot" emissions:

- emissions are usually considered cold when they are measured over a driving cycle started with the engine, and sometimes the catalyst, at ambient temperature, and hot when they are measured over a driving cycle started with the engine temperature higher than ambient temperature. We consider that these definitions lack accuracy.

- we prefer considering that hot emissions correspond only to steady emissions not affected by engine thermal condition at start; conversely cold emissions are affected by the starting vehicle temperature.

- only engine and catalyst temperatures could also be taken into account to determine the cold or hot emission character: hot emissions would thus correspond to engine thermal equilibrium. Nevertheless we consider that this is an erroneous approach since emissions reach their equilibrium before engine or catalyst temperatures.

4 - It can be therefore asserted that the first portions of NEDC and FTP75 cycles are undoubtedly cold, but do not cover the whole cold period due to cold start conditions. Thus subsequent cycles cannot be considered "hot" since they are performed partly over the cold period. In addition the third part of FTP75 is hardly cold at the beginning. Each standardized NEDC and FTP75 cycles are nevertheless ended under hot conditions and are sufficient to warm up light vehicles.

5 - Excess emissions are not systematically observed after a cold start. Indeed, at least over distances of a few kilometers, cold emission values can be lower than hot emission values. Nevertheless when considering the whole cold period, only NOx and N_2O values are decreased for non catalyst petrol and diesel vehicles; but this decrease remains relatively low, ranging from -13% to -2%.

6 - As compared to hot emissions, cold emissions of catalyst vehicles are usually higher. These relative emission can even reach very significant values for CO, HC and PAHs, reaching factors of 10, 16 and 6 respectively. In all the other cases, excess emissions range from +4% to +88%. For PAHs, excess emission rates are always high, usually by several units, but significantly vary from one PAH to the other.

7 - The distance travelled under cold conditions, i.e before emission stabilization, is relatively variable. On average, it is close to 6 km, but can be significantly higher for CH_4 and N_2O.

8 - It is widely admitted that catalysts are inefficient under cold conditions over the first kilometers travelled, even if their efficiency is satisfactory under hot engine conditions. Although the samples selected for catalyst and non catalyst vehicles slightly differ, it can be said that this is not true. In the great majority of the cases studied, under hot and cold conditions, the use of a catalyst led to a significant decrease in regulated pollutant emissions, whatever the trip duration.

9 - The taking into account of cold start conditions demonstrates that the catalyst use does not lead systematically to excess fuel consumption. An accurate evaluation should be required, integrating statistical data about vehicle usage in terms of vehicle speeds and engine temperatures.

10 - Only real cold start conditions were considered in this study, the engine being initially at ambient temperature. Under real-world conditions, the major part of starts-up are performed not long after the last stop, i.e at engine temperature between ambient and equilibrium temperatures. It would be interesting to study emission changes in these cases.

11 - Only ambient temperatures close to 15°C have been considered, i.e neither really hot, nor really cold temperatures from an European standpoint. It would be interesting to study emission changes for other ambient temperatures.

12 - Changes in particulate emissions were not studied after a cold start. This should be done considering the increasing percentage of diesel particulate emissions.

ACKNOWLEDGEMENTS

The authors wish to thank Ademe, the French Agency for Energy and Environment, for their financial support, C. Pruvost and P. Tassel from the team of INRETS emission lab, A.I. Saber and E. Combet from the Savoy University for their close co-operation in PAH analysis, as well as Mr Cupelin, from the Service de l'Ecologue Cantonal de Genève, for his assistance and the lending of test vehicles.

REFERENCES

[1] Joumard R. & M. André (1988): Real exhaust gaseous emissions and energy consumption from the passenger car fleet. SAE Congress, Detroit, USA, Oct. 31 - Nov. 3, and SAE paper 881764.

[2] Eggleston H.S., D. Gaudioso, N. Gorissen, R. Joumard, R.C. Rijkeboer, Z. Samaras & K. H. Zierock (1993): CORINAIR working group on emission factors for calculating 1990 emissions from road traffic - volume 1 : Methodology and emission factors. CEC report, Luxembourg, 116 p.

[3] André M., R. Joumard, A.J. Hickman & D. Hassel (1992) Actual car use and their operating conditions as emission parameters ; derived urban driving cycles. Higway Pollution, Madrid, 18-22 May, & Sc. Total Environ., 146/147, 1994, p. 225-233.

[4] Joumard R., R. Vidon, L. Paturel, C. Pruvost, P. Tassel, G. de Soete & A.I. Sabe (1995): Changes in pollutant emissions from passenger cars under cold start conditions. INRETS report, n° 197bis, Bron, France, 75 p.

[5] Crauser J.P. , M. Maurin. & R. Joumard (1989): Representative kinematic sequences for the road traffic in France. SAE Congress, Detroit, USA, Feb. 27 - Mar. 3, and SAE paper 890875.

[6] Prigent M. & G. de Soete (1989): Nitrous oxide N_2O in engine exhaust gases. A first appraisal of catalyst impact. SAE Congress, Detroit, USA, Feb. 27 - March 3, SAE paper 890492.

[7] Combet E., J. Jarosz, M. Martin Bouyer, L. Paturel & A. Saber (1993): Mesure par spectrofluorimétrie Shpol'skii des émissions unitaires en HAP de 30 véhicules légers et diesel selon 8 grands cycles représentatifs. 2nd. intern. symp. "Transport and Air Pollution ", Avignon, France, 10-13 Sept., 1990, and Sci. Total Environ., vol. 134, p. 147-160.

[8] Saber A., L. Paturel, J. Jarosz, J. Subtil & M. Martin-Bouyer (1993): Instrumentation for analysis of trace-level PAHs by Shpol'skii high resolution spectrofluorometry. Proceedings 13th Intern. Symp. Polynuclear Aromatic Hydrocarbons, special issue, Gordon and Breach Science Pub., Philadelpia, USA, eds P. Garrigues & M. Lamotte, p. 363-369.

[9] Fachinger C., J. Jarosz, M. Martin Bouyer, L. Paturel, A. Saber & M. Vial (1990): Microcomputer spectrofluorometric Shpol'skii data acquisition and processing. Analytica Chim. Acta, p. 203-207.

[10] CFHA (1994): Risques cancérigènes des gaz d'échappement des moteurs diesel et des moteurs à essence. Rapport de la Commission Fédérale de l'Hygiène et de l'Air, Cahiers Environ., Berne, n° 222.

[11] Lenner M. (1994): Pollutant emissions from pasenger cars - Influence of cold start, temperature and ambient humidity. VTI report, n° 400A, VTI, Sweden, 42 p.

ADDRESS OF THE MAIN AUTHOR

Dr Robert Joumard
INRETS, case 24, 69675 Bron cedex, France
tel: +33 72 36 24 77; fax: +33 72 37 68 37
email: joumard@inrets.fr

961134

Applications and Benefits of Catalytic Converter Thermal Management

Steven D. Burch, Matthew A. Keyser, Chris P. Colucci, Thomas F. Potter, and David K. Benson
National Renewable Energy Lab.

John P. Biel
Benteler Industries, Inc.

Copyright 1996 Society of Automotive Engineers, Inc.

ABSTRACT

A catalytic converter thermal management system (TMS) using variable-conductance vacuum insulation and phase-change thermal storage can maintain the converter temperature above its operating temperature for many hours, allowing most trips to begin with minimal "cold-start" emissions. The latest converter TMS prototype was tested on a Ford Taurus (3.0 liter flex-fuel engine) at Southwest Research Institute. Following a 24-hour soak, the FTP-75 emissions were 0.031, 0.13, and 0.066 g/mile for NMHC, CO, and NO_x, respectively. Tests were also run using 85% ethanol (E85), resulting in values of 0.005, 0.124, and 0.044 g/mile, and 0.005 g/mile NMOG. Compared to the baseline FTP levels, these values represent reductions of 84% to 96% for NMHC, NMOG, and CO.

BACKGROUND

Mobile sources (primarily automobiles) produce 66% of the total US toxic gas carbon monoxide (CO), 38% of non-methane hydrocarbon (NMHC) gases and 44% of oxides of nitrogen (NO_x) (contributors to smog) [Graedel, 1988]. Automobile emissions of these pollutants have decreased dramatically over the past 30 years (>95% for NMHC, on a grams-per-mile basis). However, as the number of vehicle miles traveled per year continues to increase, automakers face increasingly stringent state and federal regulations. California's ULEV (Ultra-Low Emission Vehicle) standard, which starts its phase-in during 1998, is generally viewed as the toughest to meet. It requires vehicles to produce no more than 0.040 grams of reactivity-adjusted non-methane organic compounds (NMOG) using the 1975 Federal Test Procedure (FTP-75). NMOG is similar to NMHC, but accounts for carbonyls and alcohols as well as the atmospheric reactivity of each hydrocarbon.

Today's engines use computer-controlled fuel-injection systems to minimize emission production. In addition, three-way catalytic converters are used to treat engine-out emissions. Converters are up to 97% effective at converting CO and NMHC to carbon dioxide and water vapor. However, to achieve these high conversion rates, the converters must be hot, typically about 300 C or greater. More than half (60% to 80%) of all NMHC and CO emissions occur in the first few minutes of FTP-75 while the converter is warming up [Laing, 1994]. Decreasing these "cold-start" emissions is seen as key to meeting ULEV and other future regulations.

Several approaches have been developed over the past five years for reducing cold-start emissions. Electrically heated and fuel-fired converters attempt to rapidly heat the converter at the beginning of a trip. They have been shown to reduce the warmup time from two minutes to 10 to 20 seconds, reducing CO and NMHC emissions by up to 70% [Pfalzgraf, et al., 1995]. Hydrocarbon traps attempt to capture NMHC until the converter is hot [Burk, et al., 1995]. Primary development issues for these approaches involve system cost, complexity, and durability.

Another approach is to move the converter upstream, as close to the engine manifold as possible. Close-coupled converters can reach operating temperature in under a minute, but special catalyst washcoat techniques must be used to ensure loading durability at the higher exhaust temperatures. Also, care must be taken to keep unwanted heat from the converter from flowing into the engine or passenger compartments. Despite these issues, several automakers are using or are planning to use close-coupled converters. Honda recently announced achieving ULEV standards on a Civic using this approach (and advanced engine control). Although close-coupled converters may be sufficient for some efficient four-cylinder cars (the Civic uses a 79 kW, 106 bhp engine), larger cars are likely to require additional measures.

CATALYTIC CONVERTER THERMAL MANAGEMENT

Recently there has been an effort to reduce cold-start emissions by keeping the converter hot between trips. Several automakers and EPA have evaluated adding conventional insulation to the catalytic converter [Hartsock, et al., 1994]. With fibrous refractory insulation, a converter can be kept above its operating temperature for one to two hours (versus 25 minutes for an uninsulated converter). FTP-75 cycles run one hour following a previous cycle had 80% lower NMHC emissions [Koupal, 1995].

Unfortunately, the FTP-75 test procedure requires a 12- to 36-hour cold soak between the prep cycle and measured cycle. If a thermal management system (TMS) could be developed to maintain converter temperature for 24 hours or more, automakers could achieve excellent FTP-75 emissions results. More important, overall automobile emissions could be dramatically reduced. The EPA has estimated that 98% of trips occur within 24 hours of a previous trip [U.S. EPA, 1993].

Over the past four years, a catalytic converter thermal management system has been developed by the National Renewable Energy Laboratory (NREL), a U.S. Department of Energy national laboratory. The early designs and thermal performance of this TMS are described in two previous SAE papers [Burch, et al., 1995; Burch, et al., 1994] and in a patent [Benson and Potter, 1995]. This TMS (shown in Figure 1) uses vacuum insulation to dramatically reduce heat loss from the converter shell. Metal bellows with thin porous ceramic inserts reduce conduction and radiation heat loss from the inlet and outlet.

In addition to reducing heat loss, the converter uses about 2 kg of phase-change material (PCM) to significantly boost the heat storage capacity of the converter. As shown in Figure 1, the PCM, typically a eutectic salt or metal alloy, is sealed in an annulus located between the converter monoliths and the vacuum insulation.

To protect the converter from over-heating, the vacuum insulation has a variable thermal conductance feature. This feature is described in detail in an SAE paper [Benson, et al., 1994], and a patent [Benson and Potter, 1994]. By heating a few grams of a metal hydride material containing hydrogen, the pressure of hydrogen gas within the vacuum insulation can be varied between 0.01 and 10 torr. This pressure is at most about 1% of atmospheric pressure, well below a combustion level. Even this small amount of hydrogen, however, can cause a variation in the thermal conductivity of the vacuum insulation of more than 100:1. The hydride can be actively controlled via an electric resistance heater, or passively controlled by the heat of the converter.

RECENT DESIGN IMPROVEMENTS

The converter TMS prototype reported in SAE paper 950409 demonstrated good heat retention during bench-top tests at NREL, cooling from 600 C to 300 C in just over 19 hours (see Figure 2). However, the aluminum/silicon PCM used in this prototype (PCM #1) had too high a melting point (580 C). During FTP testing at Chrysler on a Dodge Neon, the PCM could not be melted during either a standard or an extended prep cycle. Still, after a 23-hour cold soak, the converter was 146 C, hot enough to reduce the CO and NMHC emissions by 52% and 29%, to 0.27 and 0.037 g/mile, respectively (using a 11 g/l, 300 g/ft^3 palladium-only catalyst).

In response to these results, the next TMS converter prototype was designed to use a PCM with a melting point of about 350 C (PCM #2). Changes were also made to the alternating layers of copper foil and glass paper that provide radiation shielding within the vacuum insulation. The result of these changes is shown in Figure 2. The heat retention time (time from 600 C to 300 C) was increased from 19 hours to 24 hours. This figure also shows the importance of combining vacuum insulation (VCI) and PCM, versus using either alone.

The design characteristics of the present TMS converter are shown in Table 1. Prior to emission testing, the converter was aged on an engine dynamometer (Ford 5.6 l, V-8 engine). This process consisted of running the engine at a moderate speed (525 C exhaust temperature) for 24 hours, roughly equivalent to 4000 miles of durability driving. A "fully-aged" converter would have 20,000 to 50,000 miles of exposure.

Table 1 - Design Characteristics of TMS Converter

Catalyst Brick (each of 2)
Material	Cordierite	
Diameter	144 mm	5.7 in.
Length	76 mm	3.0 in.
Cell Density	62 cells/cm^2	400 cells/in^2

PM Loading
Ratio	(5:1 Pt:Rh)	
Density	2.5 g/l	70 g/ft^3

Overall Dimensions
Length	490 mm	19 in.
Diameter	216 mm	8.5 in.

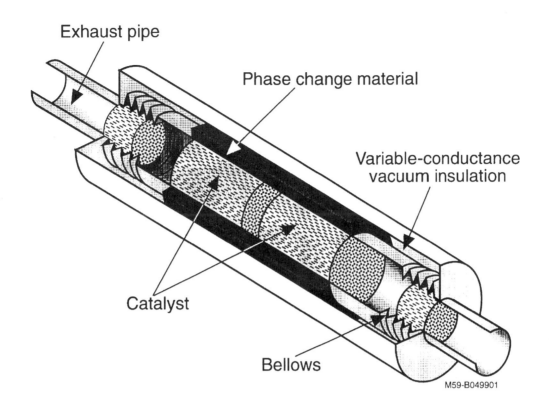

Figure 1 - Drawing of TMS converter

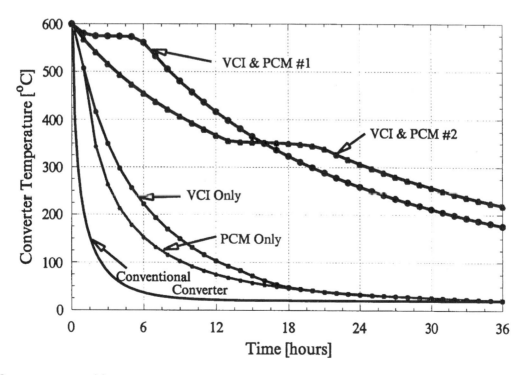

Figure 2 - TMS converter cooldown

FTP-75 EMISSIONS TESTING

After aging, the TMS converter was installed on a 1993 Ford Taurus FFV (flex-fuel vehicle) with a 3.0 liter V-6 Vulcan engine designed to run on methanol (M85) or gasoline, or any mixture of the two. The optical sensor of the test vehicle was modified to sense a combination of ethanol (E85) and gasoline instead of methanol and gasoline.

The test vehicle was also fitted with a modified exhaust system to accommodate the test article. Due to space constraints and the large diameter of the TMS converter, it was installed about 0.68 m (17") further downstream than the original (OEM) converters. Fiberglass insulation was used to reduce the heat loss in this additional pipe section (see Figure 3). An air injection port was installed upstream of the test article so that air could be injected into the exhaust stream until the engine went to closed-loop operation.

The TMS converter assembly was tested with both unleaded gasoline (Howell EEE certification fuel) and ethanol (E85). The FTP-75 uses a single Urban Dynamometer Driving Schedule (UDDS) preparation, a 18- to 36-hour ambient soak, and a three-bag FTP cycle during which HC, CH_4, CO, and NO_x are measured (NMHC = HC - CH_4).

In addition to emissions, temperatures at the following seven locations were monitored: inlet and outlet pipe, center of the front and rear bricks, between the bricks, between the front brick and intumescent mat, and between the mat and the stainless steel wall (rear brick). The pressure drop across the TMS converter was also measured. All emissions tests were performed at Southwest Research Institute.

Starting with the TMS converter at ambient temperature (25 C), at the end of the first prep cycle (a single UDDS), the minimum temperature measured in the converter was 270 C, on the steel wall around the rear brick. This indicated that although the exhaust gas was greater than 400 C for the majority of the UDDS, the PCM had not melted. Further examination of the temperature data suggested that the intumescent mats were insulating the PCM from the hot exhaust gas.

An extended prep cycle, consisting of one UDDS and multiple Highway Fuel Economy Test schedules (HFETs), was used to melt the PCM in the TMS converter. Although temperature measurements indicated that the phase change material melted after the first HFET, two HFET cycles were used for tests with gasoline and three HFET cycles for E85 tests. This procedure resulted in minimum converter temperatures of 420 C or higher for all tests, assuring that the phase-change material was fully melted.

Following an extended prep cycle, the car was "cold-soaked" (at 25 C) for 18, 24, 30, or 36 hours. After each cold soak period, an FTP cycle was performed and emissions measured. Secondary air injection was used for all tests: 2.4 l/s (5 cfm) for 110 seconds at the beginning of the test (Bag 1) and for 10 seconds following the 10-minute engine-off period of the cycle (Bag 3). Table 2 shows the total FTP emissions for various soak periods for both the gasoline and ethanol fuels.

For the ethanol fuel, exhaust samples were measured for HC, CO, NO_x, CH_4, carbonyls (aldehydes and ketones), and alcohols. Nonmethane organic gases (NMOG) were estimated using:

NMOG = (NMHC + Carbonyls + Alcohols) x RAF,

where the Reactivity Adjustment Factor, RAF = 0.67 as reported in SAE 932676 [Kroll, et al., 1993].

As the data in Table 2 indicate, by keeping the converter hot for 24 hours, the TMS substantially reduced emissions. For gasoline, NMHC and CO emissions were reduced by 84% and 91%, respectively.

The improvement with ethanol was even greater, 96% and 93%, respectively. The California Air Resource Board is discussing the possibility of creating an Equivalent Zero Emission Vehicle standard for conventional vehicles that emit no more pollution than that caused by the electric power generation used to recharge an electric vehicle. Current proposals are roughly 10% to 20% of ULEV levels. The emissions for the Ford Taurus FFV using ethanol and a TMS converter are 13%, 7%, and 22% for NMOG, CO, and NO_x, respectively.

The minimum temperature in the converter at the end of this FTP-75 test was 370 C, indicating that the phase change material had remelted. Although further testing is needed, it appears that as little as one FTP-75-type drive (11 miles) per day would keep the converter hot (between 230 C and 550 C) at all times, and emissions very low.

Maintaining the temperature within this range should also decrease thermal stress on the converter during the warmup period. Pressure drop measurements indicate the TMS converter had slightly less pressure drop than the baseline vehicle. This was also seen in a small improvement in fuel economy.

Figure 3 - Photo of TMS converter on Ford Taurus

Table 2 - FTP-75 Emissions for a Ford Taurus FFV with a Thermal Management System (TMS) Converter

	Unleaded Gasoline			Ethanol (E85)				
	Baseline	24-hour soak	30-hour soak	Baseline	18-hour soak	24-hour soak	36-hour soak	CARB ULEV
NMHC (g/mi)	0.194	0.031	0.025	0.115[a]	0.006[a]	0.005[a]	0.037[a]	-
NMOG (g/mi)	-	-	-	0.210[b]	0.006[b]	0.005[b]	0.040[b]	0.040
CO (g/mi)	1.463	0.131	0.112	1.710	0.125	0.124	0.465	1.700
NO_x (g/mi)	0.135	0.066	0.069	0.177	0.044	0.044	0.049	0.200
Converter Temperature at Test Start	25 C	233 C	189 C	25 C	290 C	235 C	175 C	-

[a] Gasoline derived NMHC = FIDHC - (CH_4 x FIDRCH4)-(Ethanol x FIDRETH);
   FIDHC - hydrocarbon measured with flame ionization detector calibrated on propane;
   FIDRCH4 - FID response factor for methane;
   FIDRETH - FID response factor for ethanol

[b] NMOG = (NMHC + Carbonyls + Alcohols) x RAF. RAF = 0.67 as reported in SAE 932676 [Kroll, et al., 1993]

CONCLUSIONS AND FUTURE WORK

Combining variable-conductance vacuum insulation and phase-change heat storage for catalytic converter thermal management is a relatively new approach to reducing cold-start emissions. Earlier papers reported on the thermal performance and some limited emission reductions, but in the present study, NMHC and CO reductions of 84% to 96% were achieved using unleaded gasoline and ethanol.

Although these results are extremely encouraging, substantial work remains before this approach can be applied to high-volume automobile production. In order to make this transition from the laboratory to the factory floor, NREL has recently teamed with Benteler Industries, Inc., a major manufacturer of exhaust system components. This lab/industry team is currently developing improved TMS converter designs which begin to address issues of cost, manufacturability, and durability.

To improve the rate of PCM melting, a metal (stainless steel) catalyst "brick" will be used. Metal monoliths are directly brazed to the steel wall, eliminating the thermal resistance of the intumescent mat. Long-term durability studies are being conducted to evaluate the containment of several potential PCMs. Other planned tests include high-temperature shock and vibration durability, emission performance with fully-aged catalysts, and verification of the hydride variable-conductance feature at extreme exhaust temperature conditions.

ACKNOWLEDGMENTS

The authors wish to thank the following sponsors and contributors to the TMS converter development and testing: Kevin Whitney and Patrick Merritt of Southwest Research Institute (emission test support), Gary Wells of Benteler Industries (gasoline testing sponsor), and Brent Bailey of NREL Alternative Fuels Program (ethanol testing sponsor, [Dodge, et al., 1995]).

REFERENCES

Benson, D.K., and Potter, T.F., 1995
U.S. Patent #5,477,676, "Method and Apparatus for Thermal Management of Vehicle Exhaust Systems," (December 26, 1995).

Benson, D.K., and Potter, T.F., 1994
U.S. Patent #5,318,108, "Gas-Controlled Dynamic Vacuum Insulation with Gas Gate," (June 7, 1994).

Benson, D.K., Potter, T.F., and Tracy, C.E., 1994
"Design of a Variable-Conductance Vacuum Insulation," SAE Technical Paper #940315.

Burch, Potter, Keyser, Brady, and Michaels, 1995
"Reducing Cold-Start Emissions by Catalytic Converter Thermal Management," SAE Technical Paper #950409.

Burch, S., Keyser, M., Potter, T., Benson, D., 1994
"Thermal Analysis and Testing of a Vacuum Insulated Catalytic Converter," SAE Technical Paper #941998.

Burk, P., et al., 1995
"Cold Start Hydrocarbon Emissions Control," SAE Technical Paper #950410.

Dodge, L., et al., 1995
"Development of a Dedicated Ethanol Ultra-Low Emission Vehicle - Phase 2 Report," NREL Technical Paper 425-8195, NTIS #DE95013145, September 1995.

Graedel, T., et al., 1988
"Ambient Levels of Anthropogenic Emissions and their Atmospheric Transformation Products," in Air Pollution, the Automobile and Public Health, A.Y. Watson, National Academic Press, Washington, pp. 133-160.

Hartsock, D., Stiles, E., Bable, W., Kranig, J., 1994
"Analytical and Experimental Evaluation of a Thermally Insulated Automotive Exhaust System," SAE Technical Paper #940312.

Koupal, J., 1995
"Controlling Emissions Following Intermediate Soak Periods," SAE Cold Start Emissions TOPTEC, Jan. 12-13, 1995, Golden, CO.

Kroll, M., et al., 1993
"Influence of Fuel Composition on NMOG Emissions and Ozone Forming Potential,' SAE Technical Paper #932676.

Laing, P.M., 1994
"Development of an Alternator-Powered Electrically-Heated Catalyst System," SAE Technical Paper #941042.

Pfalzgraf, B., et al., 1995
"The System Development of Electrically Heated Catalyst (EHC) for the LEV and EU-III Legislation," SAE Technical Paper #951072.

U.S. Environmental Protection Agency (EPA), 1993
"Federal Test Procedure Review Project: Preliminary Technical Report," EPA 420-R-93-007, May 1993.

961137

Electrically Heated Catalyst - Design and Operation Requirements

F. Terres, H. Weltens, and D. Froese
Heinrich Gillet GmbH & Co. KG

Copyright 1996 Society of Automotive Engineers, Inc.

ABSTRACT

EHC design and engine operation requirements for a battery powered EHC-cascade were investigated using flow rig, engine dynamometer and vehicle evaluations.

Low mass and Pd-coated heater elements and light-off converters are recommended for optimum light-off performance. Raising the heating power improves light-off. However, battery powered systems are limited to 1.5 kW.

Rich engine operation combined with an excess of secondary air results in high exothermic energy output. The benefit of additional heating and the impact of cascade position (close coupled or underfloor) are closely related to the test cycle.

ULEV limits were achieved using a MY 91 vehicle without upgrades in engine control.

INTRODUCTION

New automobile catalytic converters attain maximum conversion rates of about 99% under optimum operating conditions. However, this requires an exhaust gas temperature of at least 350°C at the catalytic converter inlet. The present test regulations for measuring automobile emissions prescribe several hours of car conditioning at an ambient temperature of 20 to 30°C, preceeding each emission test. The test cycles in the USA (FTP 75) and Europe (MVEG) begin with an idling phase and moderate acceleration. Under such conditions, at least 120s elapse before the underfloor catalytic converter attains its light-off temperature. During these first two minutes, the exhaust gas is incompletely cleaned. Hence, more than 80% of the unburnt hydrocarbons (HC), of a 20 to 30 minute exhaust gas test, are emitted during this initial period.

Emission limits have become more stringent in the 1990's. New techniques were developed to accelerate the warm-up of the catalytic converter. The objective is to attain the light-off temperature within 20s following a cold start. The light-off temperature is generally defined as the exhaust gas temperature, at the inlet to the catalytic converter, at which a HC conversion of 50% is achieved. Retarded ignition, fuel enrichment and secondary air, supplied to the exhaust gas system is one of the basic techniques to reduce HC and CO emissions during the warm-up phase. Other methods practiced are: fabricated exhaust manifolds in sheet metal, thin-wall air gap insulated pipes between the engine and the underfloor catalytic converter, close coupled catalytic converters with high temperature resistant coatings, and by-pass systems.

Further reduction in pollutant emissions is possible by active systems. These deliver additional heat to the catalytic converter after a cold start, reducing the light-off time. Examples of such concepts are: gasoline burner systems to heat the exhaust gas [1], exhaust gas ignition by operating with extremely enriched air-fuel mixtures combined with secondary air supply [2], and electrically heated catalytic converters (EHC) [3,4,5,6,7,8]. The potential of these concepts was investigated in view of the EURO III (from 1999) and ULEV emission standards. The vehicular weight and the raw engine emissions are decisive factors for the choice of the concept. The raw emissions can be drastically reduced with specific measures in the intake and fuel system and through optimization of the combustion process. In the future, selective exhaust gas sensors will substantially contribute to the optimization of the combustion process. These and other engine sensors deliver high frequency data to a high-speed microprocessor. It computes and controls the optimum air/fuel ratio, ignition timing and secondary air supply.

The application and optimization of an EHC system are described below. The investigations encompassed many of the EHC-designs available from catalytic converter substrate manufacturers internationally

FUNCTION OF A CASCADE EHC SYSTEM

This section describes the optimization of the design and operating parameters of the cascade EHC system.

The EHC cascade consists of the heater element (igniter), the light-off converter (LOC) and the main catalytic converter (MC). The heater element should initiate the CO and HC oxidation in the subsequent LOC, as quickly as possible. This exothermic reaction heats the exhaust gas. Together with the enthalpy of the exhaust gas, and to a limited extent the additional electrical energy, the subsequent MC is very quickly brought to its operating temperature. It is highly advisable to compactly accommodate all components in a single housing. This housing should be thermally insulated. The cascade of active and passive catalytic converters can be located underfloor or near the engine.

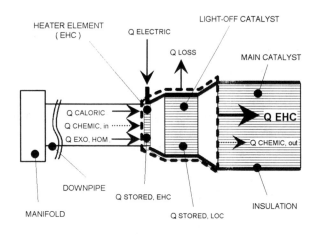

Fig. 1: Energy balance of an EHC cascade

Figure 1 shows the energy balance in an EHC cascade. The objective is to rapidly maximize the heat Q_{EHC} released from the heater and pre-converter to the main catalytic converter. This heat is:

$$Q_{EHC} = Q_{CALORIC} + Q_{EXOTHERMIC} + Q_{ELECTRIC}$$

with

$Q_{CALORIC}$ =
 combustion heat of the engine out gases reduced by the heat retained by manifold, downpipe and LOC ($Q_{STORED,LOC}$), and further reduced by heat loss to the ambient (Q_{LOSS}).

$Q_{EXOTHERMIC}$ =
 energy released through homogenous gas reactions in the manifold and downpipe as well as through catalytically supported reactions in the EHC and the LOC.

$Q_{ELECTRIC}$ =
 electrical energy supplied to the EHC reduced by the energy retained within the heater element.($Q_{STORED,EHC}$)

The released enthalpy first heats the LOC. After the light-off temperature is exceeded in the LOC, further energy is released to heat the main catalytic converter. The MC more rapidly attains its operating temperature, mainly due to the energy released by the homogeneous and catalytically activated gas reaction.

DESIGN AND OPERATING PARAMETERS OF THE EHC SYSTEM

An extensive concept study was performed to identify the parameters influencing the enthalpy sources. The following design and operating parameters were investigated and optimized.

Design parameters
- Design of the heater element (metal foil matrix or extruded matrix).
- Mass and total surface area (TSA) of the heater element.
- Catalyst coating on the heater element.
- Distance between the heater and the LOC.
- Mass, volume and material (metallic foil or ceramic support) of the LOC.
- Catalyst coating and loading on the LOC

Operating parameters
- Electrical power supply and heating time.
- Engine management (air/fuel ratio and ignition timing).
- Secondary air management (quantity and injection time).

This is a large number of relevant parameters. Hence, individual investigations were performed on a flow rig, engine dynamometer and chassis dynamometer. Some parameters were investigated in parallel.

4. BASIC INVESTIGATIONS AND DESIGN OPTIMIZATION ON THE FLOW RIG

Figure 2 is a schematic of the cascade EHC system investigated in the flow lab. Propane (C_3H_8) is mixed with cold air supplied at a constant flow rate of 15 kg/h. For different tests the combustible gas mixture had a fixed propane content between 100 and 10,000 ppm. A propane concentration of 5,000 ppm corresponds to an exothermic potential of about 1.6 kW. This is similar to the exothermic potential under engine idling conditions. The heater element ignites the propane-air mixture. The electrical output of the heater was varied between a minimum of 1.5kW and a maximum of 2.0 kW. Tests on the flow rig confirmed that raising the power from 1.5 to 2 kW results in a considerable improvement of light-off performance. However, a maximum rating of 1.5 kW is considered reasonable without enhancing the vehicular electrical system, e.g. an upsized alternator or an additional battery.

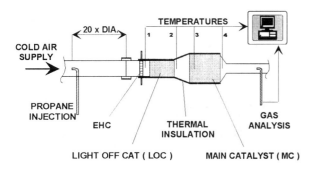

Fig. 2: Principle of EHC efficiency test rig

The pertinent gas temperatures were measured at locations 1 to 4. Further, the HC concentration was continuously registered downstream of the main catalytic converter. The design parameters studied were the different EHC concepts supplied by different EHC manufacturers, the influence of catalytically coating the EHC, the distance between the EHC and LOC, and the LOC volume. Altogether, 4 different EHC designs were analyzed: two extruded elements and two metal foil elements.

The main advantage of the investigations in the flow laboratory is the excellent reproducibility of the results. The EHC cascade has a modular design. Thus, rapid and easy modification of the individual components is possible. Further, the operating parameters can be quickly and easily modified. The measurements are very precise. The composition of the synthetic exhaust gas, including the relatively large air surplus, are not completely representative for the exhaust gas of a spark ignition engine. Nevertheless, the results are valuable and permit accurate comparison of the individual cascade designs and the operating parameters.

INFLUENCE OF EHC IMPREGNATION - Figure 3 shows the influence of catalytically coating the heater element. It charts the propane conversion efficiency as a function of elapsed time after switching on the heater element. For these tests, no main converter was used, to accentuate the effect of catalyst coating on the heater element. One heater element was coated with a Pt/Rh catalyst, the other heater element was only wash-coated to account for the influence of the wash-coating on the thermal response. Both elements were measured individually as well as in combination with an LOC. The propane conversion curve clearly shows the effectiveness of the catalytic coating. The enthalpy released during the catalytic reaction in the EHC heats the downstream LOC, thus starting its conversion substantially sooner. Therefore, catalytically coating the heater element is highly recommended. The additional energy necessary to heat the coating is more than offset by the exotheric energy generated in the catalyzed heater.

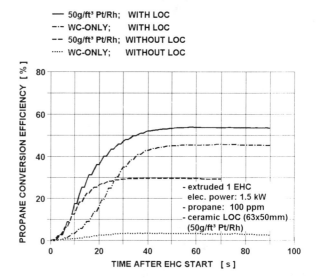

Fig. 3: Impact of heater impregnation on light-off performance (without main converter)

COMPARISON OF DIFFERENT HEATER ELEMENT DESIGNS - The basic function of the EHC is to initiate the chemical reaction in the downstream LOC. Since exothermic reactions transfer more heat to the gas than can be provided via electrical energy, it is imperative to reach the activation temperature of the EHC as soon as possible. This can be achieved by low mass heater elements. Also, in the heater, low total surface area accelerates light-off through reduced heat transfer to the gas flow. The EHC module in the cascade was exchanged to compare different EHC designs. The investigation included two metal foil EHCs and two extruded EHCs, all from different manufacturers. The LOC and the main catalytic converter remain constant in this experimental series. Figure 4 is a direct comparison of the different EHCs. As expected, larger EHC mass and bigger total surface area noticeably delay activation of the HC conversion. The thermographs (Appendix), taken after 5s and 15s when heating with 1.5 kW at an air flow rate of 15 kg/h, explain the observed differences in light-off performance. Optimum heat up is achieved with low surface area heater elements, irrespective of whether metal foil or extruded element are used. Metal foil II EHC reacts too slow due to the high total surface area. It should be mentioned, however, that the prototype used in this investigation was not the most recent design.

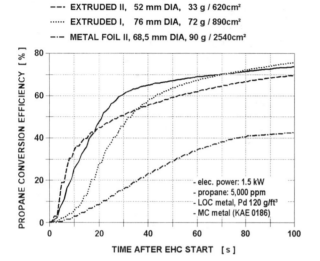

Fig. 4: Impact of EHC mass and geometry on light-off performance

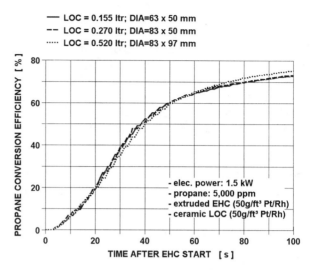

Fig. 5: Impact of LOC volume on light-off performance

Since the heater element by itself has a low flow restriction, it will result in a relatively bad flow distribution when it is attached to an inlet pipe with a smaller diameter. Thus a heater element diameter in the range of the inlet pipe diameter is recommended. This compact design also fulfills the requirement of a low mass EHC.

INFLUENCE OF THE LIGHT-OFF CONVERTER VOLUME AND MASS - When considering constant cell density and wall thickness, the volume of the LOC has no impact on the HC conversion efficiency measured downstream of the MC (figure 5). Even shortening the LOC length by 1/3 does not essentially impact the conversion. The conversion contribution of the missing LOC volume is compensated for by the main converter. However, removing the LOC totally has been found to deteriorate light-off performance. Hence, the LOC volume should be minimized for reasons of economy and compact construction.

Further tests indicated that decreasing the substrate wall thickness improves light-off. Thus compact, thin-wall and low mass LOCs are recommended.

DISTANCE BETWEEN HEATER AND LIGHT-OFF CATALYTIC CONVERTER - Figure 6 shows that a gap of 20mm between the heater element and the LOC retards the light-off. This can be explained by the heat loss due to heat absorption by the housing in the region of the gap.

However, a gap between the EHC and the LOC is necessary when using a LOC with an increased diameter. A bigger LOC diameter is useful to reduce the diameter step between the small EHC and the main converter, for which a maximum diameter to length ratio is recommended to achieve low exhaust system back pressure. Further, a bigger LOC diameter will also reduce the contribution of the LOC to overall back pressure.

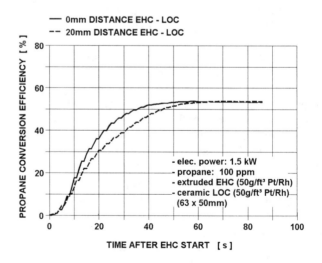

Fig. 6: Impact of distance between EHC and LOC on light-off performance (without main converter)

The extruded EHC is usually supplied as an individual element, allowing a combination with a LOC of larger diameter. When a metal foil substrate is used, the EHC and the supporting LOC are supplied as a single module with nearly the same diameter. The best developed metal foil EHC has a gap of about 5mm between heater element and LOC. This small gap improves light-off as shown in figure 6.

PARAMETER STUDY ON ENGINE TEST BENCH

A number of operating parameters of an EHC cascade were investigated on an engine test rig. The tests were performed under realistic but steady-state boundary conditions. Figure 7 is a schematic of the engine test rig.

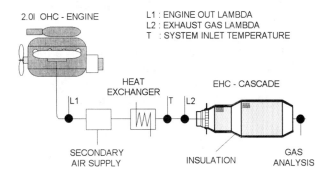

Fig. 7: Principle of engine bench setup

The engine was operated at constant conditions: 0.82 engine out lambda and an idling speed of 1030 RPM. The exhaust gas flow rate was 18 kg/h. The HC concentration in the exhaust gas was 1000 ppm C_3H_8, giving a possible exothermic energy of the unburned HC of about 400W. The CO concentration is 2% and represents an exothermic energy potential of 1200W. Thus the exhaust gas contains a total energy potential of 1.6 kW. Under constant operating conditions, the exhaust gas lambda at the EHC was varied between 0.82 and 1.2 through secondary air supply. The experiments, intended to investigate the operating parameters, were performed at an exhaust gas temperature of 40°C at the inlet to the EHC cascade. A heat exchanger kept this exhaust gas temperature constant during the entire investigation. In practice, the exhaust gas temperature continuously increases after a cold start and contributes to heat up the LOC. Thus, compared to real exhaust conditions, this simulation is more stringent. It accentuates the differences observed for different operating parameters. The EHC was heated during 20s for all experiments. This corresponds to the idling period in the FTP test.

For all tests, done on the engine bench, the same 1,24l ceramic main converter (1.77g/l Pt/Rh) was used in combination with various heater elements and LOCs.

ELECTRICAL HEATING - Particularly in the first seconds after an engine start, the power input decisively influences the response of the EHC cascade. Test results demonstrated the substantial light-off improvement when raising the EHC power. However, a battery powered system is limited to 1.5 kW.

During acceleration (in the FTP cycle after 20s), the rapidly increasing exhaust gas flow can hardly be heated with a "reasonable" electrical input. Heating of the EHC should therefore be limited to the idling phase of the exhaust gas test.

SECONDARY AIR STRATEGY - Figure 8 displays the influence of the secondary air. The results show that the exhaust gas lambda must have a minimum value of at least 1.0 to permit high exothermic reactions. Further increase of secondary air supply only yields a relatively small improvement in the HC conversion. However, it is seen that when exhaust gas lambda exceeds 1.0, the EHC cascade is relatively insensitive to the secondary air quantity. Complex and expensive control of the secondary air injection rate is therefore not necessary in practice.

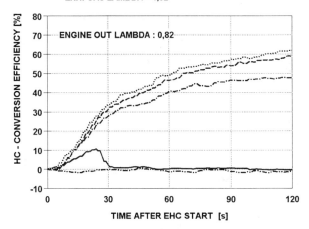

Fig. 8: HC-conversion efficiency as a function of the amount of secondary air supply

EFECT OF AIR/FUEL RATIO - Running the engine rich results in a substantial increase of engine out CO emission. The HC concentration only increases slightly. Together with sufficient secondary air, supplied to the exhaust gas, the higher CO content results in a considerably higher exothermic heat and thus in an improved light-off performance. As indicated in figure 9, the accumulated HC emission is considerably lower due to the light-off improvement although the engine out HC emissions were slightly higher. The break even point for the accumulated CO emission was found here after 90s. Generally, in view of future limits, the CO emission is not critical. However, the cold start enrichment has to be properly adjusted in terms of the CO limit.

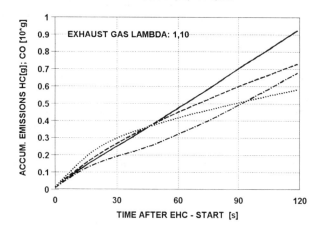

Fig. 9: Impact of engine-out lambda on HC and CO emissions

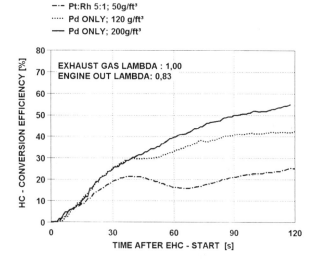

Fig.10: HC-conversion as a function of coating (EHC and LOC both metal foil)

IMPACT OF IMPREGNATION - The CO content of the exhaust gas is substantially higher than the HC content. The reaction enthalpy of conversion is mainly from the CO. Therefore, a coating should be chosen that has a good CO conversion at low temperatures. The influence of the coating is seen in Figure 10. Obviously, the usual Pt/Rh impregnation is not optimum for light-off performance during the warm-up phase. The Pd-only impregnation is more favorable. For the presented exhaust gas lambda of 1.0, the 7.06g/l (200 g/ft^3) impregnation is superior to 4.24g/l (120g/ft^3). However, further test results have shown that at higher exhaust gas lambda (e.g. λ=1.1) both loadings perform similarly.

INVESTIGATION OF THE EHC CASCADE ON THE VEHICLE DYNAMOMETER

Figure 11 is a schematic of the vehicle emission test setup. The EHC cascade is located underfloor. The distance between the exhaust manifold and converter inlet is 900 mm. The vehicle is equipped with a thin wall tubular manifold. The wall thickness of the single skin downpipe is 1.0 mm. Both downpipe and manifold are insulated with an external 6 mm thick ceramic wool insulation.

For all tests an EHC cascade was used, consisting of a metal foil heater element combined with a metallic LOC and a ceramic main converter. EHC and LOC are impregnated with 4.23g/l Pd-only. The main converter has a 1.77g/l Pt/Rh impregnation.

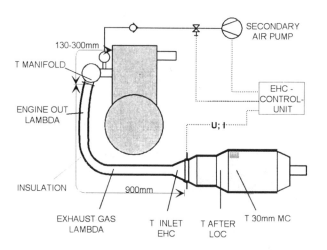

Fig.11: Vehicle emission test setup with EHC system (1.6l CVH engine; MY91)

SECONDARY AIR STRATEGY - Figure 12 shows the accumulated HC emissions during the first 140s of the FTP-75 cycle for various durations of secondary air supply. Engine out lambda was only slightly rich (λ~0.95). The EHC was heated during the first 20s with 1.6 kW.

With increasing duration of secondary air supply, the HC emissions are significantly reduced. An HC emission reduction of 60% was observed, when air is supplied during the entire interval between engine start and begin of lambda control, which was 35s for this test vehicle.

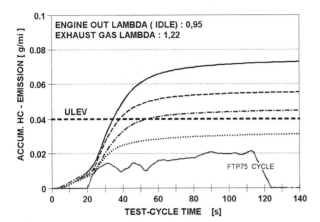

Fig.12: Accumulated HC emissions in the FTP75 cycle vs secondary air injection time

Longer open loop operation and secondary air supply might result in further HC and CO emission reduction. However, the NOx emission is then becoming critical since the NOx conversion efficiency of a catalytic converter is very poor under lean engine operation.

With further reduction in the engine out lambda during cold start, the HC and CO emission reduction is expected to be even more significant.

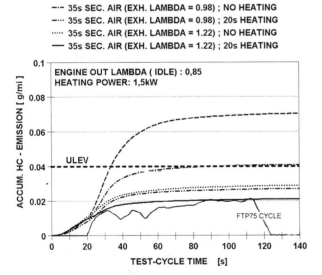

Fig.13: HC emissions for different cold start strategies in the first 140s of FTP75 cycle

FTP RESULTS - The impact of different secondary air strategies, with and without heating, is shown in Figure 13. At 140s after a cold start, the accumulated HC emission is reduced by 60% due only to the secondary air supply. Additional heating results in a further HC emission reduction of only 10%.

EHC mass and electrical power rating are decisive factors for achieving EHC operating temperature rapidly. Since the test vehicle is already equipped with a mass optimized EHC, further improvement can only be expected by raising EHC power rating.

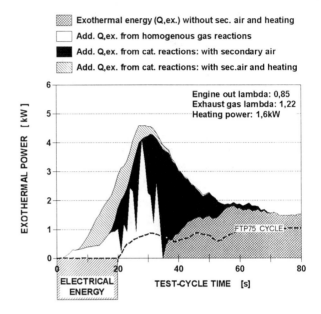

Fig.14: Exothermic heat transfer to the main converter in the FTP75 emission test

The exothermic energies for the different cold start strategies are presented in Figure 14. They indicate the light-off improvement due to homogenous gas reactions and earlier converter activation.

Supplying secondary air significantly increases the exothermic energy during the first 50 sec of FTP75 test, compared to the baseline test (no air, no heat). The conversion in the catalytic converter commences after 18s. Additional heating activates the converter already 8s after test start. That results in a surplus of exothermic energy. However after about 35s, the benefit of heating is insignificant compared to secondary air only.

Comparing the energies obtained using secondary air supply with the surplus when heating, underlines the dominant effect of secondary air strategy.

As shown in figure 15, the exhaust gas temperature after the LOC exceeds the gas temperature at the cascade's inlet due to the exothermal reactions in the EHC and LOC. When only injecting secondary air, the cross over point appears after 18s. Through additional heating, this point was only marginally advanced to 14s.

American FTP-75 test. Hence, the influence of EHC heating would be expected to be more significant. In view of stage III limits, the first 40 s idling are eliminated. EHC power was supplied for 20s. Secondary air injection was limited to 25s, where the vehicle begins its first deceleration.

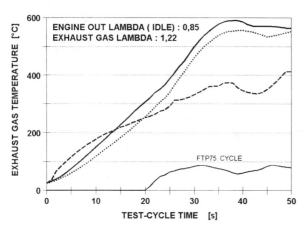

Fig.15: Exhaust gas temperatures during FTP75 emission test

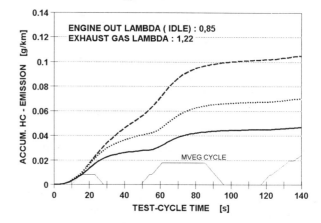

Fig.17: HC emissions for different cold start strategies in the first 140s of MVEG cycle (40s idling eliminated)

Figure 17 shows that HC emissions were reduced about 30% by secondary air injection, combined with cold start enrichment and retarded ignition. A further reduction of 25% was attained through electrical EHC heating.

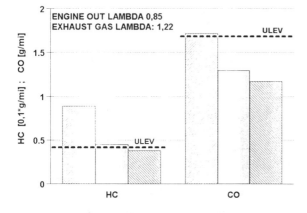

Fig.16: FTP75 emission test results for different cold start strategies

The total FTP75 test results, presented in figure16, demonstrate that emissions in the range of ULEV were achieved. It is remarkable that these good results were obtained primarily using a combination of conventional passive means like low thermal inertia manifold and downpipe, retarded ignition, cold start enrichment and secondary air injection. The effect of heating was only marginal. It should also be mentioned, that the test vehicle didn't represent the latest technical status. Newer engines would be expected to produce even better results.

MVEG RESULTS - The driving speeds in the European MVEG test are much lower than in the

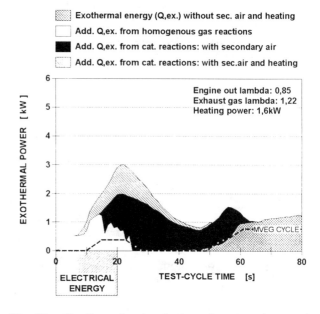

Fig.18: Exothermic heat transfer to the main converter during MVEG emission test

The heat, transferred to the MC through exothermic reactions is plotted in Figure 18. The exhaust gas temperature is low at the EHC cascade inlet. Hence, there is relatively little heat contributed by the homogenous gas reaction in the exhaust manifold and in the downpipe, compared to the FTP-75 test. The contribution of the heterogeneous catalytic reactions to the heat transfer to the MC is dominant. When the EHC heating is switched on, the catalytically supported HC and CO oxidation occurs only 7s after a cold start. In contrast, the catalytic reactions in the EHC and LOC without heating only occur after 15s.

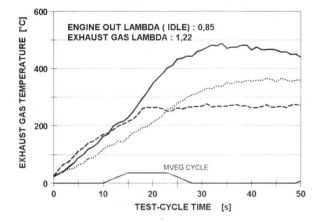

Fig.19: Exhaust gas temperatures in the MVEG emission test cycle

The exhaust gas temperatures, plotted in figure 19 demonstrate again the benefical effect of heating in the MVEG cycle. The temperature after the LOC without EHC heating exceeds the inlet temperature after about 23s, whereas with heating this occurs after only 11s.

This positive effect is also seen in Figure 20. A considerable reduction of emissions is noticeable for both HC and CO.

SUMMARY AND CONCLUSIONS

Different electrically heated catalytic converters were investigated regarding their potential to meet future emission limits. Two metal foil heater elements and two extruded EHC, all supplied by different manufacturers, were investigated.

The different EHC were combined with a light-off converter (LOC) and a main converter (MC), to form a modular EHC cascade.

Fundamental conceptual investigations were performed on a flow rig and on an engine bench. The

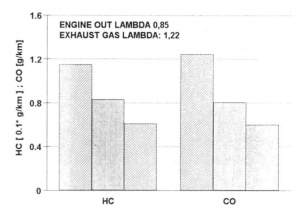

Fig.20: MVEG emission test results for different cold start strategies

investigations of the EHC cascade were performed in both cases under steady-state operating conditions. The different EHC cascade concepts were verified based on FTP75 and MVEG emission tests with a 1.6 liter vehicle on the chassis dynamometer. The power rating of the EHC was restricted to 1.6 kW to avoid major expensive modification of the vehicle's electrical system. Secondary air was supplied during the first 35s (FTP75) or 25s (MVEG) after a cold start. Close loop operation starts 35s after cold start. Thermal inertia was reduced by equipping the vehicle with a sheet metal manifold and a thin-wall downpipe.

The important results are summarized as follows:

> The EHC functions as an igniter. It promotes and accelerates the catalytically supported HC and CO reactions. The EHC is less effective in heating the exhaust gases through convection or radiation.
> The MC is heated through exothermic homogenous gas reactions in the exhaust manifold and in the downpipe, as well as through catalytic supported reactions in the EHC and LOC.
> A compact and low-mass EHC combined with a low-mass LOC with a large total surface area is advantageous
> Optimized concepts with extruded or metal foil EHC have almost equivalent performance.
> The housing of the EHC cascade should be thermally insulated to reduce heat loss to ambient. Further, the distance between the EHC and LOC should be minimized. Stepped diameters, increasing from EHC to MC, improve flow distribution in the individual elements and reduce pressure loss.

> The main energy source is the CO and not the HC. Considerable cold start enrichment is necessary to raise the energy potential of the exhaust gas.
> An exhaust gas lambda in the range 1.1 to 1.2 is required. Because of the relatively large optimum range, expensive secondary air control is not necessary.
> In the FTP-75 test: cold start enrichment, retarded ignition and secondary air supply already achieve a 60% reduction of the HC emission. Additional EHC heating only yields a further 10% improvement, for a total reduction of 70%.
> In the MVEG test cycle, the same concept attains an HC reduction of only 30% with air but without EHC heating and approx. 55% with electrical heating. The approx. 25% contribution of the EHC heating to the HC reduction is therefore significant.

Overall, a battery-powered EHC system can be very effective both for HC and for CO reduction. However, in the FTP-75 test, the effectiveness of an electrically heated catalyst is disproportionately small, considering the technical effort. Here, more efficient alternatives are close coupled main converters, or underbody converters (UBC) in combination with close coupled light-off converters and thin-wall, air-gap insulated pipes.

Under European test conditions, the potential of the battery-powered EHC system on emission reduction is more significant. However, most of the vehicles seem to meet stage III limits with passive technology. Hence the application of an EHC system will be a niche solution.

ACKNOWLEDGMENT

The authors wish to thank Mr. Brück (Emitec), Dr. Montierth (Corning), Mr. Bates (NGK) and Mr. Lylykangas (Kemira) for the helpful discussions during the evaluation of these investigations.

REFERENCES

[1] Kollmann, K. et al., "Concepts for Ultra Low Emission Vehicles", SAE Paper 940469

[2] Ma, T. et al., "Exhaust Gas Ignition (EGI) - A New Concept for Rapid Light-Off of Automotive Exhaust Catalys", SAE Paper 920400

[3] Breuer, J. et al., "Electrically Heated Catalyst for Future USA and European Legislation", SAE Paper 960339

[4] Socha, Louis S. et al., "Emissions Performance of Extruded Electrically Heated Catalysts in Several Vehicle Applications", SAE Paper 950405

[5] Kaiser, F.W. et al., "Optimization of an Electrically-Heated Catalytic Converter System Calculations and Application", SAE Paper 940465

[6] Pfalzgraf, B. et al. "The System Development of Electrically Heated Catalyst (EHC) for the LEV and EU-III Legislation, SAE Paper 951072

[7] Shimasaki, Yuichi et al., "Study on Conformity Technology with ULEV Using EHC System", SAE Paper 960342

[8] Abe, F. et al., "An Extruded Electrically Heated Catalyst: From Design Concept through Proven-Durability", SAE Paper 960340

APPENDIX 1

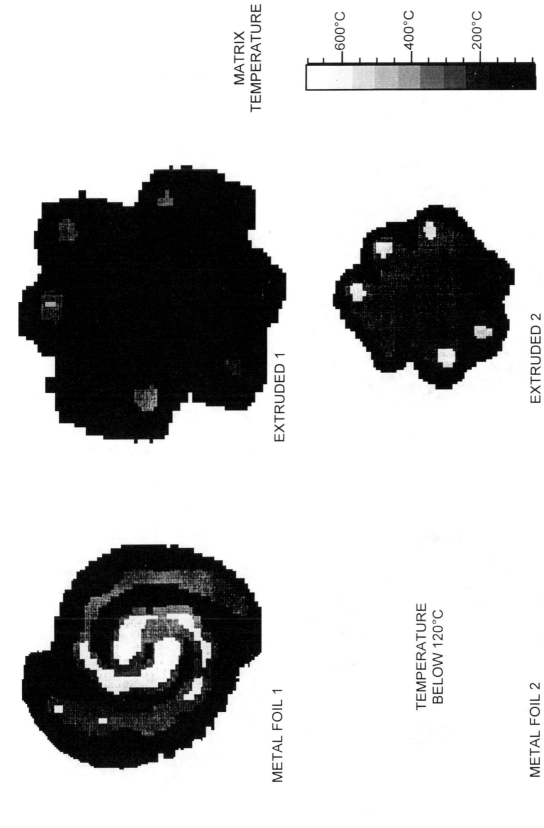

EHC - MATRIX TEMPERATURE AFTER 5s HEATING AT AIR MASS FLOW RATE OF 15 kg/h AT 20°C

APPENDIX 2

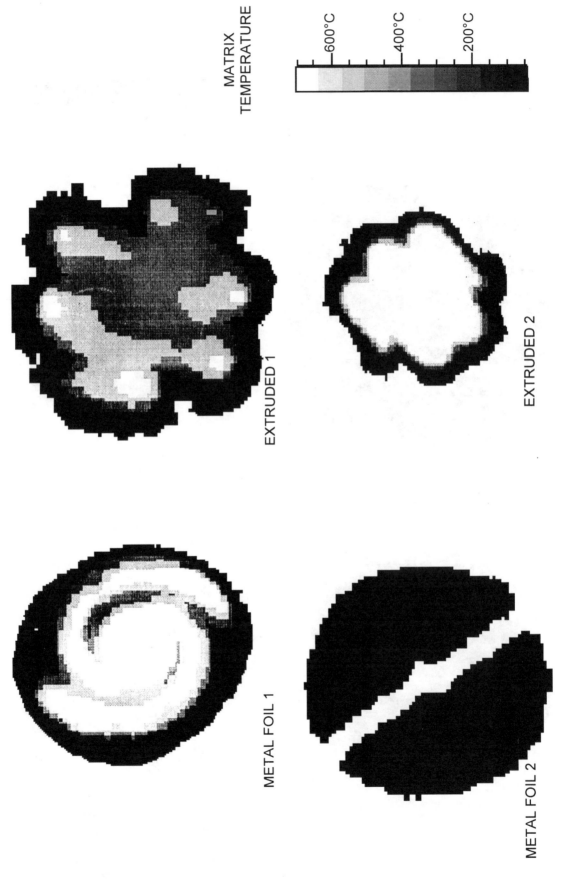

EHC - MATRIX TEMPERATURE AFTER 15s HEATING AT AIR MASS FLOW RATE OF 15 kg/h AT 20°C